The Myth of
Mirror Neurons

The Myth of Mirror Neurons

The Real Neuroscience

of

Communication and Cognition

•———◆———•

Gregory Hickok, PhD

W. W. NORTON & COMPANY

NEW YORK • LONDON

Copyright © 2014 Gregory Hickok

For information about permission to reproduce selections from this
book, write to Permissions, W. W. Norton & Company, Inc.,
500 Fifth Avenue, New York, NY 10110

For information about special discounts for bulk purchases, please
contact W. W. Norton Special Sales at specialsales@wwnorton.com
or 800-233-4830

Manufacturing by Courier Westford
Book design by Brooke Koven
Production manager: Louise Parasmo

Library of Congress Cataloging-in-Publication Data

Hickok, Gregory.
The myth of mirror neurons : the real neuroscience of communica-
tion and cognition / Gregory Hickok, PhD. — First edition.
pages cm
Includes bibliographical references and index.
ISBN 978-0-393-08961-5 (hardcover)
1. Mirror neurons. 2. Brain—Physiology. 3. Cognitive neuroscience.
I. Title.
QP376.H53 2014
612.8′233—dc23

2014011426

W. W. Norton & Company, Inc.
500 Fifth Avenue, New York, NY 10110
www.wwnorton.com

W. W. Norton & Company Ltd.
Castle House, 75/76 Wells Street, London W1T 3QT

1 2 3 4 5 6 7 8 9 0

To Krissy, for 30 years of
encouragement, support, and love

Contents

The Myth of
Mirror Neurons

Preface

A Neural Blueprint for Human Behavior?

THE DISCOVERY and characterization of DNA in 1953[1] changed biology forever. DNA is the blueprint for life, the key to understanding how organisms are built, how they evolve, and how things can go wrong in disease.

In 2000, psychologist V.S. Ramachandran invoked the epochal impact of DNA in a prediction regarding a then recently discovered class of brain cells, mirror neurons:

> I predict that mirror neurons will do for psychology what DNA did for biology: they will provide a unifying framework and help explain a host of mental abilities that have hitherto remained mysterious and inaccessible to experiments.[2]

(Judging from the title of Ramachandran's TED talk on mirror neurons delivered in 2010, "The Neurons that Shaped Civilization," a decade of intense research on the cells seems to have confirmed his predictions.)

In 2008, Marco Iacoboni, a neuroscientist at UCLA, echoed Ramachandran's enthusiasm for this class of brain cells:

We achieve our very subtle understanding of other people thanks to certain collections of special cells in the brain called mirror neurons. These are the tiny miracles that get us through the day. They are at the heart of how we navigate through our lives. They bind us with each other, mentally and emotionally. . . . Mirror neurons undoubtedly provide, for the first time in history, a plausible neurophysiological explanation for complex forms of social cognition and interaction.[3]

Ramachandran and Iacoboni are not alone in their excitement. The scientific journals are full of mirror neuron–based explanations of human language, imitation, social perception, empathy, mind reading, music appreciation, art appreciation, spectator sports enjoyment, stuttering, and autism. A *Wall Street Journal* article in 2005 explained "How Mirror Neurons Help Us to Empathize, Really Feel Others' Pain"; in the same year *NOVA* aired a program simply titled "Mirror Neurons," and in 2006 *The New York Times* reported on them in an article titled "Cells that Read Minds." Books and blogs tout the virtues of mirror neurons for everything from classroom education to your golf game. The extent of a human male's erection has been claimed to be related to mirror neuron activity.[4] If a recent news story is correct, even the Dalai Lama was enticed to visit UCLA to hear about the role of mirror neurons in cultivating compassion. Fitting for a class of cells sometimes called "Dalai Lama neurons."

What are these miraculous human brain cells that can explain everything from erections to autism? Curiously, all of this speculation about human behavior does not find its foundation in *human* neuroscience research at all. Instead, the theoretical keystone is a class of cells found in the motor cortex of pigtail macaque monkeys, animals that can't talk, don't appreciate music, and, frankly, aren't all that nice to each other. The behavior of mirror neurons is modest, at least in the context of the human abilities they are claimed to enable. The fundamental feature of these cells is that they respond ("fire," as neuroscientists say) both when a monkey reaches for an object and when the monkey watches someone else reach for an object. That's about it. From this simple response pattern neuroscientists and psychologists have built one of the most wide-reaching theories in the history of psychology for the neural basis of human behavior.

What is it about this apparently simple response pattern of mirror neurons in macaques that has excited an entire generation of scientists? How is it possible that a cell in the motor cortex of a *monkey* can provide the neural blueprint for human language, empathy, autism, and more? What are the logical building blocks that allow this inference?

The basic idea is simple—that's its appeal. When a monkey reaches for an object, it "understands" its own action, what the goal is, why that particular goal is targeted, and so on. In short, the monkey "knows" what it's doing—and why. That's the trivial part. But what enquiring monkey minds *really* want to know is what *other* monkeys are up to. *Is he going to try to steal my food or is he just on his way to the watering hole?* That's rather harder to figure out. So the question is, how do you read (or understand) the actions of others? Mirror neurons provide a simple answer because they fire *both* when the monkey executes an action *and* when it observes similar actions executed by other monkeys: if the monkey understands the meaning of its own actions, then by *simulating* the actions of others in its own neuronal action system, it can, by the same token, understand the meaning of the actions of others.

It's an ingenious trick—using knowledge of self-actions to collect intelligence on the intentions of others—with potential applications that extend well beyond the monkey lab. Given this starting point, the inferential steps from mirror neurons to human communication and cognition are not hard to imagine. Humans too need to read the intentions of other humans' actions, so maybe they have a mirror system as well. Speech is an important human action; maybe a mirror neuron–based simulation mechanism underpins the crown jewel of human cognition: language. And sport is nothing if not action based; maybe we get so fanatical about our home team because mirror neurons put us right on the field, simulating every throw, catch, and kick. With humans it's not just actions that we are capable of understanding. We understand others' emotions and mental states; maybe there is a mirror-like mechanism behind empathy. Some disorders, like autism, are popularly thought to involve a lack of empathy; maybe autism results from broken mirror neurons. And mirror neuron theorizing comes with an evolutionary kicker: by showing a (previously missing) link between a monkey neuron that is involved in recognizing the actions of others and high-level human cognitive abilities, mirror neurons provide a toehold for a theory of the evolution of the human mind. Example: if mirror neurons in monkeys enable the understanding of simple gestural actions, such as grasping an object, then to start down the evolutionary path toward language all natural selection needs to do is hit on a way of broadening the scope of action understanding such that it includes actions related to vocalization, like monkey calls. The explanatory power of these deceptively simple cells appears impressive indeed.

Like many cognitive neuroscientists, I was intrigued by mirror neurons when I first learned about them in the 1990s. But I didn't get seriously interested in their properties and the hypotheses that swirled around them until the theoretical whirlwind kicked up dust in my own area of research. I study the neural basis of speech and language with a particular focus on speech perception and sensorimotor functions: How does the brain take a variable stream of air pressure waves (what hits your ears when you listen to speech) and convert it into recognizable sounds, words, sentences, and ideas? How do we learn to articulate the sounds of our language? Why do we talk with an accent when we learn a second language? Why does an echo of our own voice in a bad phone connection make it so hard to talk? Where does our "inner voice" come from—the sense that we can hear our voice when we "talk in our heads"—and what purpose does it serve? These are some of the questions I was (and remain) interested in answering.

Mirror neurons, according to a growing chorus of scientists, answered all these questions and more. Before long, I found myself fielding questions about mirror neurons in my own research presentations at conferences and such. *Don't mirror neurons explain that? That can't be right because we know mirror neurons are the basis of . . .* Although I was deeply suspicious that a cell type in a macaque brain could explain human language, I couldn't ignore mirror neurons any longer.

In the spring of 2008 I organized and taught a graduate course on these monkey cells. The study plan was simple. Each week my students and I read a set of original mirror neuron research papers and then came to class ready to discuss and debate. To broaden the conversation I took the course online via *Talking Brains*, a blog I had started two years prior with my longtime collaborator David Poeppel. I used the blog to post the reading list, summarize the course discussions, and host an international conversation on the topic.

I was surprised by what I learned reading through those papers. The claims just didn't seem to add up. Looking back, one particular comment that I posted on the blog after a classroom discussion of a seminal mirror neuron review paper[5] summed up my impression of the entire enterprise:

> Quite an interesting read, but I have to say, it's not exactly the tightest paper I've read. Plenty of speculation, hints of circularity, over-generalization, etc. We all agree that mirror neurons are very interesting neural creatures, but the idea that they are the basis for action understanding is darn near incoherent.

Having schooled myself on the basics of mirror neurons and having formed a skeptical perspective, I took my thoughts on the road. Now in my presentations I would devote 10 minutes or so pointing out why mirror neurons didn't solve all of the problems I was interested in. It was intended as theoretical self-defense. But in the audience at one of these talks sat the editor of the *Journal of Cognitive Neuroscience*, where several mirror neuron–related papers had been published. He asked me if I would be interested in writing a critique of mirror neurons for publication in his journal. I agreed to the task and in 2009, despite a fair amount of resistance from one anonymous peer reviewer, the editor decided the work should be published. The article was titled "Eight Problems for the Mirror Neuron Theory of Action Understanding in Monkeys and Humans," and it has received a fair amount of attention. As of this writing, it is the journal's most downloaded.

Many of the problems that I pointed out in that paper appear in various places and in various contexts in the chapters to follow:

- There is no direct evidence in monkeys that mirror neurons support action understanding (Chapter 3).
- Mirror neurons are not needed for action understanding (Chapters 4 and 6).
- Macaque mirror neurons and mirror-like brain responses in humans are different (Chapter 3).
- Action execution and action understanding dissociate in humans (Chapters 4–8).
- Damage to the hypothesized human mirror system does not cause action understanding deficits (Chapters 4–6).

Having jumped headfirst into the debate, I started doing my own research aimed at testing the mirror neuron theory within its flagship human application—and my research expertise—language (Chapters 5 and 6). The theory hasn't held up well. For example, in one large-scale study of speech recognition involving over 100 individuals with brain injuries (mostly due to stroke) we found that deficits in understanding speech were associated with damage to auditory-related brain regions rather than the supposed mirror system (Chapter 5).

I also continued doing my homework on mirror neuron applications beyond the language domain, including work on imitation (Chapter 8), autism (Chapter 9), and empathy (Chapters 2, 8 & 9), and I continued to blog about what I learned.

As you may have gathered by now, much of my writing on mirror neurons has been critical, pointing out the many ways that the theory falls short on logical or empirical grounds. One can get a less-than-favorable reputation with this approach. As Sam Rayburn, former speaker of the US House of Representatives, once said, "A jackass can kick a barn down, but it takes a carpenter to build one." I suspect that many in the mirror neuron camp, and perhaps beyond, class my commentary with that of Rayburn's jackass. After all, in science, like barn work, it is easier to kick down than to build and I'll be the first to admit that I've done my fair share of kicking. In my defense, those of us who have serious questions about mirror neuron claims are up against more than a mere barn. By the mid-2000s, the theory had morphed from "speculation" (to use the mirror neuron discoverers' own word) to a virtual theoretical fortress. When faced with a juggernaut of this sort, the only way to promote alternative ideas is to shine as bright a light as possible on the weaknesses of the theory. My "Eight Problems" paper aimed to do exactly this and the present book toes the same line, to some extent.

But I have some carpentry skills too, or at least I can recognize a well-built barn when I see one. Accordingly, *The Myth of Mirror Neurons* isn't just a barn-kicking exercise. The chapters that follow contain plenty of discussion of alternative explanations for what mirror neurons are really doing (Chapter 8) and how communication and cognition work in the brain (Chapters 5–10).

When I got involved, there were few skeptics (at least public skeptics). For example, the Wikipedia entry on mirror neurons circa 2010 contained all of three sentences summarizing my "eight problems" critique under a "criticism" subheading. It's not so much that mine was the only dissenting voice, just that I was probably the most blatant "jackass." Today's Wikipedia entry includes a subheading titled "Doubts concerning mirror neurons" that is seven paragraphs long and summarizes the critiques of several prominent scientists in a range of disciplines. The international debate over mirror neurons and indeed the nature of human cognition has intensified. Mirror neurons are no longer the rock stars of neuroscience and psychology that they once were and, in my view, a more complex and interesting story is gaining favor regarding the neuroscience of communication and cognition. But I'm getting a bit ahead of the story.

I

Serendipity in Parma

I N 1988 a group of neuroscientists in Parma, Italy, headed by Giacomo Rizzolatti, began work on an experiment aimed at trying to understand how neurons in the macaque monkey premotor cortex control grasping actions (see Appendix A for a primer on brain organization and terminology). The team employed the "single unit recording" method in which microelectrode probes inserted in the brain measure electrical activity of single neurons ("units") while the animal is performing a task. Specifically, the microelectrodes pick up *action potentials*, also called "spikes," which are the electrical blips that a neuron generates to communicate with other neurons. We can think of action potentials as the basic element in the language of neurons and single unit recording as a way of wiretapping that conversation. The method is tried and true—research using single unit recording has resulted in more than one Nobel Prize[1]—and is often regarded as the gold standard in animal-based neurophysiological research. It is fallible, however, particularly in terms of interpreting the content of the neural code, what the firing patterns tell us about the cell's information processing function.

By the time the Parma team started their experiment, work in the same lab already indicated that cells in a particular brain region in the macaque called F5 (frontal lobe area 5; see the nearby figure) fired

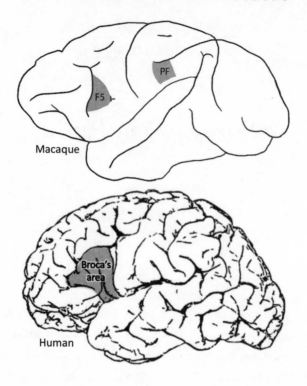

Macaque

Human

more vigorously when the monkey executed grasping actions, often in a grip-specific manner: some cells responded more for a "precision grip" (e.g., a thumb–index finger posture that you would use to grasp a raisin) while others fired more for a whole–hand grip (e.g., grasping an orange), and so on. Thus, the cells appeared to code specific grip types that the monkey may need for grasping objects. Some preliminary data also suggested that some of these F5 grasping neurons fired, in addition, when the monkey merely *saw* graspable objects.

Interestingly, the cells' "preferred" grip type during grasping (i.e., that which caused the most firing) and their preferred object type during viewing tended to match; for example, precision grip neurons responded only when a small object was presented. This suggested that the neuron firing patterns were coding the relation between object shape and the grip types they called for. Given this observation, the working hypothesis about neurons in F5 was that these cells as a group represented a kind of vocabulary of possible motor actions and that input from the sensory features of objects selected appropriate actions from this motor vocabulary.[2] For instance, viewing a tennis ball–sized object would activate neurons that code a whole-hand grip, while viewing a raisin–sized object would activate neurons that code a precision grip.

The idea that F5 neurons might be coordinating visual information about objects with motor programs for grasping was interesting and important, but more data were needed to confirm the hypothesis. In particular, the response of these grasping cells to visually presented object information was not thoroughly characterized. One practical complication was how to tease apart a cell's response to seeing the object from its response associated with the immediately following grasping action. Because the two events—seeing and grasping—partially overlapped in time, it was impossible to determine which event was responsible for changes in the cell's firing pattern. The team designed an experiment to get around this problem.[3]

As researchers recorded the activity of individual neurons, a single monkey performed the following task seated in front of a box that had a one-way mirror forming a door on the front. The monkey was trained to press and hold a switch that lit the inside of the box to reveal an object of a specific size and shape. After a delay of one to two seconds, the door opened and the monkey could release the switch, reach into the box, pick up the object, and score a morsel of food hidden underneath. If the monkey released the switch before the programmed delay ended, the door would not open. This design ensured that researchers could measure the neural response to seeing the object during the delay and cleanly separate it from the response associated with the reaching movement once the door opened.

The experiment ultimately resulted in a finding that had nothing much to do with the study's original design. In between trials, the experimenter would reach into the box, swap out the objects and replace the food morsel while the monkey sat there and watched, waiting for the next opportunity to perform the task. But even while the experiment was in intermission, the microelectrodes in the monkey's brain were still recording neural activity. During these in between times, the experimenters noticed something. A sizable proportion of the cells they were monitoring seemed to respond to the reaching and grasping actions of the *experimenter* as he set up the next trial. Previous studies had reported neurons that responded to observed actions, but these neurons were in sensory areas, never in a motor area.[4] A more formal experiment aimed at recording the response to *action observation* in the monkey's brain was quickly set up.

After measuring a neuron's response in the grasping task according to the design of the original study, the experimenters used themselves as the stimulus, performing a series of actions in front of the monkey. Some actions involved a food morsel, including putting it on a surface, picking it up, giving it to a second experimenter, and then stealing it away. Other objects were broken, folded, or torn. The experimenters waved their hands, lifted their arms, made threatening gestures. They also pantomimed object-directed actions (such as grasping an invisible raisin) or grasped objects with a tool instead of their hand.

When the dust settled, the team had recorded data from 184 neurons. Almost all of the recorded neurons responded when the monkey performed some kind of action, consistent with F5 being a motor area. But 87 of these neurons also responded to some type of visual stimulus. As found previously, most of the visual responses (48) were to simple objects, but 39 responded to the observation of actions. In 12 of the 39, researchers observed a correspondence between the cell's preferred action type in observation and execution: if the cell responded most strongly when the monkey executed a grasp with the hand, it also responded most strongly when observing the same action, a grasp with the hand. The remaining 27 action-responding cells were less selective in terms of observation-execution correspondence. For example, the cell might respond to the observation of a hand *placing* an object on the table and during the execution of *grasping* an object or bringing an object to the mouth. These cells showed a kind of logical relation between observation and execution preferences as if they coded something like, *you place a raisin, I grasp it*. Eleven cells showed this type of pattern. None of the 39 cells that fired during action observation were found to be responsive to the other types of actions (pantomimes, grasping with a tool, etc.) performed for the monkey.

The team published a short paper on the findings in 1992.[5] In it the authors first proposed an explanation for why some cells in motor cortex might respond to actions:

> One of the fundamental functions of the premotor cortex is that of retrieving appropriate motor acts in response to sensory stimuli. If one considers the rich social interactions within a monkey group, the understanding by a monkey of actions performed by other monkeys must be a very important factor in determining action selection . . . . [this idea] fits well in the conceptual framework of current theory on the function of premotor cortex. (p. 179)

I couldn't agree more. Just like object size and shape are critical during action selection—for example, whether to use a pincher or whole-hand grip—so too, the actions of other animals are important for action selection. Therefore, it makes sense that the brain has a mechanism for communicating information about perceived actions of others to motor areas that code self-actions for execution.

If the report had ended with this interpretation, the finding would still be interesting and important for the scientific community, but it would never have evolved into a theory that made newspaper headlines. However, the last paragraph of the report took up the 12 cells that showed a close correspondence between the actions they preferred in observation and execution.

The authors pointed to two areas of human research that seemed to show a similar observation-execution correspondence. One was limb apraxia, a condition in which affected individuals have difficulty *executing* voluntary movements. It had been reported that some patients with limb apraxia also had deficits in *understanding* the limb gestures of others.[6] The second area of human research the authors invoked was speech. Here they referred to the motor theory of speech perception, a theory developed in the 1950s by Alvin Liberman and colleagues at the Haskins Laboratories in New Haven, Connecticut. The basic idea is that we perceive speech not by reconstructing in our brain the speech sounds themselves but instead by reconstructing the motor gestures that generated those sounds (Chapter 5 provides more details on this theory). The motor theory of speech perception had spawned decades of research, much of which failed to provide support for the hypothesis, and so by the time the Parma team began their recordings, speech scientists had largely abandoned the motor theory. Nonetheless, the motor theory of speech perception, like the limb apraxia reports, claimed that perception of actions (including speech actions) depended on the motor system. The Parma team concluded their report:

> Although our observations by no means prove motor theories of perception, nevertheless they indicate that in the premotor cortical areas there are neurons which are endowed with the properties that such theories require. It is interesting to note that the anatomical location of . . . F5, corresponds in large part to that of Broca's area in the human brain. (p. 179)

The tie-in with Broca's area (see earlier figure) was critical. While the details of its function remain murky, Broca's area has long been implicated as a motor speech area. It's one of the two classic language areas in the brain and has been a central hub in models of language-brain relations since the 1860s when French neurologist Paul Broca first claimed that it was the seat of articulate speech.[7] As we see in the quote above, the monkey equivalent of Broca's area, anatomically speaking, is widely considered to be area F5. Before the discovery of mirror neurons this was largely inconsequential because even though F5 and Broca's area might be anatomically related, there was no real functional connection according to standard assumptions: Broca's is a speech area and monkeys don't talk. But now, in light of the existence of mirror neurons in F5, the Parma researchers suggested such a functional link. The implication was that human language has its evolutionary roots in F5 monkey-see-monkey-do cells, as mirror neurons were called briefly.[8] This substantially upped the theoretical ante because it opened the door to explaining aspects of human cognition in terms of neural mechanisms that are present in monkeys.

The 1992 report made no mention of the term "mirror neurons" and scientists working in the broader fields of neuroscience and psychology were for the most part unaware of the discovery of mirror neurons as well as the authors' profound conjecture for much of the next decade.

Rizzolatti's group published a more thorough paper on this new class of cells in the journal *Brain* in 1996.[9] This paper, titled "Action Recognition in the Premotor Cortex," reports basically the same experiment but involves two monkeys and recordings from 532 neurons in area F5. The findings from the 1992 paper were largely reproduced. Cells were found in area F5 that responded both when the monkeys grasped objects and when the animals observed an experimenter performing the same or similar actions. The authors coined the term *mirror neurons* to refer to this class of cells.

The 1996 *Brain* report showed that the initial observations in one monkey were not an accident. The finding replicated. The report also made a stronger case for the relevance of mirror neurons to human behavior. The authors asked the question, *Does a mirror system exist in humans?* And their answer was effectively, *yes.* Three lines of evidence in support of this conclusion were discussed, two from their own group and one from the extant literature:

1. *A transcranial magnetic stimulation (TMS) study showed that when humans observe grasping actions, their own hand-related motor excitability is increased.*[10] TMS is a noninvasive technique in which strong magnetic fields are passed through the skull and focused on underlying brain regions. (See Appendix B for details on cognitive neuroscience methods.) The magnetic fields affect the electrical properties of the neurons in the stimulated area. Using different stimulation protocols, TMS can be used either to excite neurons or to suppress or interfere with their activity. In a 1995 study by the Parma team, TMS was used to excite neural activity in primary motor cortex regions controlling

the hand and arm. Stimulation of these areas results in involuntary twitches of the hand and arm muscles. The amplitude of these twitches can be recorded with electrodes attached to the skin lying over the muscles of the hand; these are called motor-evoked potentials (MEPs). The study found that the MEPs were larger when people were observing object-directed grasping actions performed by others compared to when they observed the objects alone. The study authors interpreted this as evidence that observing actions automatically activates the grasping-related motor system.

2. *A positron emission tomography (PET) study showed that when humans observe actions performed by others, this results in an increase in neural activity in Broca's area, the region thought to be the human homologue of macaque area F5.*[11] PET is a so-called functional brain imaging method (see Appendix B), which means simply that it records brain activity rather than making static images of brain structure. PET records brain activity indirectly by measuring regional changes in blood flow. When a particular brain area is active, blood flow to that region increases to deliver the metabolic fuel, oxygen and glucose, needed to support the additional activity. Although rather indirect reflections of the activity of neurons, techniques that measure blood flow—such as PET and its more recently developed cousin, functional MRI (fMRI)—have proven to be invaluable to human neuroscience. The finding by Rizzolatti's group that blood flow increases in Broca's area during action observation was consistent with the idea that the human motor system is engaged during action perception.

3. *Patients with frontal lobe lesions in the general vicinity of Broca's area have been reported to have pantomime recognition deficits.* Neurological literature documents that patients with language disorders (aphasias) sometimes have deficits in recognizing pantomimes, identifying the intended action in a pantomime gesture. This is

clearly relevant to the question of what mirror neurons might be doing because pantomime recognition can be construed as a form of action understanding. Although the consensus around 1996 was that the damage that caused such deficits did not involve Broca's area, in light of the PET study of action observation, the Parma team suggested instead that it was indeed damage to Broca's area that caused the pantomime recognition deficits.

Armed with data that confirmed the existence of mirror neurons in monkeys and with three suggestive lines of evidence for a similar system in humans, Rizzolatti and colleagues came to the conclusion in their 1996 *Brain* paper that not only does a human mirror system exist, but that it critically involves Broca's area, the human equivalent of monkey area F5, and is involved in action recognition: "the different lines of evidence reviewed above appear to indicate that both F5 and Broca's area have a hand movement representation and that, probably, they are both endowed with a similar mechanism for action recognition" (p. 607).

As in the 1992 paper, the research team closed with an even more strongly worded speculation regarding the role of mirror neurons in human speech: "considering the homology between monkey F5 and human Broca's area, one is tempted to speculate that neurons with properties similar to that of monkey 'mirror neurons', but coding phonetic gestures, should exist in human Broca's area and should represent the neurophysiological substrate for speech perception."

Some years later, in 2002, mirror neurons were reported outside of F5 in a sensorimotor region of the monkey parietal lobe called PF (previous figure). Had mirror neurons in the parietal lobe been discovered first, the tie-in to speech may not have been so obvious or influential. But it didn't play out that way and by the time the parietal mirror neurons were reported, the proposed link among monkey mirror neurons in F5, human Broca's area, and speech had permeated the field. The effect was profound. If monkey mirror neurons could crack the case on how the brain processes speech, they might be capable of opening the door to understanding a wider range of human behaviors. Research and theorizing on the scope of the human applications of these "tiny miracles," to borrow Iacoboni's term, exploded in the late 1990s and for the next decade. As the next chapter shows, it really did seem like monkey mirror neurons would unlock the mysteries of the *human* mind.

2

Like What DNA Did for Biology

MIRROR NEURONS UNLOCK THE SECRETS OF LANGUAGE,
MIND READING, EMPATHY, AND AUTISM

I FIRST HEARD about mirror neurons in a presentation by Giacomo Rizzolatti delivered in 1999 at the annual meeting of the Cognitive Neuroscience Society in San Francisco. By that time, the potential implications of mirror neurons were beginning to creep into the consciousness of neuroscientists and psychologists. I recall seeing videos from the original experiments during Rizzolatti's talk and thinking that the cells were pretty interesting. But I didn't give mirror neurons much more thought for several years. Despite speculation about the role of mirror neurons in my area of neuroscientific research, speech and language, I just didn't see the relevance at the time. It was well known that damage to frontal lobe speech-related motor systems, including Broca's area, the region where mirror neurons were supposed to live, did not cause deficits in speech recognition, a function they were supposed to support. Obviously, I reasoned, the cells weren't all that important for speech.

I later ran into Rizzolatti at a meeting of the Society for Neuroscience and asked him directly about the "language problem" for his theory of mirror neuron function. His theory claimed that the ability to perceive speech sounds—to recognize or "understand" the difference between, say, *cat* and *pat*—depended on the human mirror system, Broca's area in particular. (In the context of human research, the term *mirror system* is often used in place of *mirror neurons* because the behavior of individual cells was not directly observable in humans; Chapter 3 contains the details on this issue.) I expressed my concern that his prediction doesn't hold up: we have known for decades that damage to Broca's area and surrounding regions can impair speech production, but largely spares the ability to understand speech. So maybe it's true that monkey mirror neurons are involved in action understanding, but there is no way that this is the case for speech, I argued. He agreed with my characterization of the data, but disagreed with my conclusion. He explained to me that the mirror system was complex and involved a network of several interacting regions that would be hard to disrupt with a single focal region of damage. Later, while teaching my course on mirror neurons I came across the exact same argument in a 2004 review paper on mirror neurons written by Rizzolatti and collaborator Laila Craighero:[1]

> At first glance, the simplest, and most direct, way to prove that the mirror-neuron system underlies action understanding is to destroy it and examine the lesion effect on the monkey's capacity to recognize actions made by other monkeys. In practice, this is not so. First, the mirror-neuron system is bilateral [involves both hemispheres] and includes, as shown above, large portions of the parietal and premotor cortex. Second, there are other mechanisms that may mediate action recognition. . . . Third, vast lesions as those required to destroy the mirror neuron system may produce more general cognitive deficits that would render difficult the interpretation of the results. (p. 173)

Despite this kind of counterargument, I remained unconvinced. The evidence from language was just too powerful to be dismissed. After all, damage to Broca's area and surrounding regions *does* cause speech *production* deficits—the human speech production system is not bilateral, like the monkey mirror system, according to Rizzolatti—and if this speech "action execution" system is important for action understanding, why don't stroke victims with damage to this system have speech understanding deficits? It didn't add up in my mind and so, despite claims to the contrary, I concluded that mirror neuron work simply did not apply to my area of research.

Apparently my view was a minority one. Starting in the late 1990s others saw profound implications for mirror neuron research, not only for speech and language, but also for social cognition (mental functions that enable social and cultural interaction). The swell of excitement followed the publication of two theoretical papers in 1998. One coauthored by mirror neuron discoverer Rizzolatti and USC computational neuroscientist Michael Arbib, titled "Language within Our Grasp," made the argument that mirror neurons may unlock the secrets of language.[2] The other, a collaboration between Vittorio Gallese, a codiscoverer of mirror neurons, and philosopher Alvin Goldman, titled "Mirror Neurons and the Simulation Theory of Mind-Reading" proposed that mirror neurons provided the foundation for the human ability to read the mind of others.[3]

The concept of "mind reading" is not as far-fetched as it sounds. We do it every day. If you see a man standing at the door of his car with one hand clinging to a bag of groceries and the other hand frantically digging through his pockets, you infer immediately what he is thinking and feeling, and maybe even what he might do next. This is the mind reading that Gallese and Goldman referred to and it falls within a domain of social cognitive research dubbed "theory of mind," the ability to consciously recognize one's own mental states—"I'm tired," "I'm frustrated," "I want a hamburger"—and to realize that other people have their own distinct mental states that can be different from our own.

These two papers represented a turning point in mirror neuron research. They tackled, head on, two of the human mind's most sophisticated and socially relevant abilities, language and mind reading (theory of mind), and linked them to an apparently simple neural mechanism already present, presumably, in the common ancestor between humans and monkeys. Suddenly, mirror neurons seemed to unlock the mysteries of the evolution of the human mind. Accordingly, these two papers received a lot of attention in broad scientific circles and remain two of the most highly cited papers in all of psychology and neuroscience in the last decade and a half. After they were published, the number of scientific papers appearing *each year* that used the term "mirror neuron" in the title or abstract began a steep upward trajectory, roughly doubling in number every two years between 2000 (4 papers) and 2010 (135 papers).

The two 1998 articles made very similar claims, arguing that mirror neurons were the evolutionary precursor for the two respective human abilities. And both claims were built on the assumption that mirror neurons comprised a mechanism for action recognition, that is, that we understand actions by *simulating* others' actions in our own motor system. To quote from Rizzolatti and Arbib:

> Our proposal is that the development of the human lateral speech circuit is a consequence of the fact that the precursor of Broca's area was endowed, before speech appearance, with a mechanism for recognizing actions made by others. This mechanism was the neural prerequisite for the development of interindividual communication and finally of speech. (p. 190)

And Gallese and Goldberg:

> Our speculative suggestion is that a "cognitive continuity" exists within the domain of intentional-state attribution [inferring others' intentions, mind reading] from non-human primates to humans, and that [mirror neurons] represent its neural correlate. . . . The capacity to understand action goals, already present in non-human primates, relies on a process that matches the observed behavior to the action plans of the observer. . . . Action-goal understanding . . . constitutes a necessary . . . stage within the evolutionary path leading to the fully developed abilities of human beings. (p. 500)

The logic seems straightforward. If mirror neurons allow the monkey to recognize the action of fellow monkeys (or their experimenters) by simulating those actions, then it is just another evolutionary step or two for mirror neuron *circuits* (neurospeak for a neural system that performs some function) to do the same in the linguistic or mind reading domains. Language: I know how to speak and what I mean when I speak; therefore I understand the speaking actions of others by simulating their speech in my own brain. Mind reading: I know that I have a mind that can represent different mental states (thoughts, desires, beliefs); therefore, I understand what other people are thinking by simulating their situation in my own mind. *Simulation*, or *observation-execution matching*, explained how these abilities were accomplished and mirror neurons provided the neural mechanism.

This is the context for V. S. Ramachandran's now famous prediction that the discovery of mirror neurons will do for psychology and neuroscience what DNA did for biology. Indeed, as predicted, more mental abilities were soon explained with a mirror simulation mechanism. In fact, in 2001—the year after Ramachandran's essay appeared at Edge.org—two new publications attributed even greater explanatory power to mirror neurons. One was Gallese's "The 'Shared Manifold' Hypothesis: From Mirror Neurons to Empathy," which generalized the mirror neuron simulation model to a proposed explanation of empathy: "the neural matching mechanism constituted by mirror neurons—or by equivalent neurons in humans—... is crucial to establish an empathic link between different individuals."[4] The notion draws on our sense that empathy involves putting oneself in another's shoes emotionally, or *simulating* the feelings of another. The other publication came from a group in the UK who were not involved in the discovery of mirror neurons but saw a link to their own work. The team, led by autism researcher Justin Williams, authored a paper titled "Imitation, Mirror Neurons and Autism,"[5] in which they proposed that mirror neuron dysfunction could explain autism. The assertion was that several of the claimed characteristics of autism—difficulty taking others' perspectives, lack of empathy, deficit in imitation ability, language problems, and so on—can be viewed as deficits of mental simulation. Since mirror neurons support mental simulation, dysfunction of the mirror system must lie at the core of autism. (Chapter 9 discusses this idea at length.)

Step back a bit and it might seem implausible that the behavior of a class of neurons in the motor cortex of a monkey could unlock the secrets of the human mind. But if you start with the assumption that mirror neurons in macaques support action understanding in that primate (a theoretical possibility) and that a similar class of neurons exists in humans (fairly weak *direct* evidence at present but highly probable in my view), then the extrapolation of mirror neuron function to human abilities like language, mind reading, and empathy is a reasonably straightforward step. And even more straightforwardly, autism, which involves presumed deficits in language, mind reading, and empathy, is then neatly explained as a dysfunction of the mirror system. These were legitimate and thoughtful scientific arguments and were taken very seriously by the scientific community.

While research continued intensely on the core theoretical applications of mirror neurons (action understanding, speech and language, mind reading, empathy, autism), the excitement spread throughout the scientific community. Before long, mirror neurons were implicated in any number of human abilities, phenomena, and disorders, some reasonable, some less so. Everyone, it seemed, was jumping on the bandwagon. In a stream of publications in scientific journals over the next decade mirror neurons were implicated in

- Lip reading[6]
- Stuttering[7]
- Schizophrenia[8]
- Comprehension of action words[9]
- Imitation[10]
- Phantom limbs[11]
- Neurorehabilitation[12]
- Hypnosis[13]
- Misattribution of anger in the music of avant-garde jazz saxophonists[14]

- Sexual orientation[15]
- Cigarette smoking[16]
- Music appreciation[17]
- Political attitudes[18]
- Felt presence[19]
- Facial emotional recognition[20]
- Obesity[21]
- The degree of male erection[22]
- Psychopathic personality disorder[23]
- Love[24]
- Contagious yawning[25]
- Business leadership[26]
- Self-awareness of emotional states[27]
- "Our aesthetic response to art, music, and literature, the dynamics of spectatorship, and resistance to totalitarian mass movements," to quote from one report[28]
- Mother–infant communication and emotion processing[29]
- Perception of vocally communicated emotion[30]
- Spectator sport appreciation[31]
- Development of Jungian collective unconscious archetypes and self-agency[32]
- Mass hysteria[33]
- Drug abuse[34]
- Favoritism[35]
- Mother–infant attachment[36]
- Efficacy of group psychotherapy[37]
- Risk assessment[38]
- Walking dreams in congenitally paraplegic individuals[39]
- Pain synesthesia[40]
- Self-awareness in dolphins[41]

And these examples are drawn only from publications with "mirror neurons" in the title or abstract and indexed in *PubMed*, a database of scientific publications developed by the National Center for Biotechnology Information at the US National Library of Medicine.

If you simply Google "mirror neurons," here are some headlines and quotes you'll find:

- Learn how to use mirror neurons to influence people
- Barack Obama is tapping into your brain [via mirror neurons]
- Mirror neurons—what some dub "Dalai Lama neurons" —force us to rethink the deepest aspects of our very selves
- The neurons that shaped civilization
- Sex = mirror neurons!
- Prayer and mirror neurons
- Gay mirror neurons come out
- God created human beings with these mirror neurons so that "do unto others" would make sense to us all
- Education and parenting are among some of the human endeavors most reliant on the proper functioning of mirror neurons
- How Pinterest profits from neural mirroring

Mirror neuron lore has clearly spread well beyond the lab and into our daily lives.

Before we get too excited, it's important to recognize that all of these theoretical generalizations of mirror neuron function are grounded in two assumptions. The first is the claim that mirror neurons in macaque monkeys are the basis of that species' ability to understand actions. The second is the idea that humans also have a mirror system that serves the same function. These assumptions are codependent. In fact, as we see in the next chapter, each leans on the other in a rather unhealthy way, especially given that if the first assumption is incorrect, then the logic behind mirror neurons as the basis for human speech, mind reading, empathy, and so on simply falls apart.

3

Human See, Human Do?

RESEARCH ON the human mirror neuron system has been aimed at determining whether in fact mirror neurons exist in humans, and if so, what they are doing. Direct evidence of their existence is thin but ultimately this doesn't concern me much. To my mind, it is virtually a given that humans have mirror neurons. How can I be so confident? Because humans can easily imitate observed actions (see Chapter 8) and to do this, there must be a neural link between an observed action and an executed action of our own; that is, *some* system must transform observed actions into executed actions.

his Q

For me, the real question is the second: what is our system doing?

mirror

EARLY INVESTIGATIONS INTO THE HUMAN MIRROR SYSTEM

RECALL THAT the earliest preliminary publication on monkey mirror neurons in 1992 speculated about a role for mirror neurons in human behavior, speech perception in particular. This suggestion sparked an immediate, intense effort to identify mirror neurons in humans and to understand their functional properties (under what conditions they activate), an effort that continues in earnest today. But there's a methodological challenge with this research program. Mirror neurons, by definition, are cells that respond both when the animal performs an action and when it views the same or similar action. So to identify a mirror neuron, you need to directly record a single cell's activity by using an electrode poked into brain tissue but you can't do this in humans, except in exceptional medical circumstances. As a result, most research on the human mirror system has had to settle for indirect methods, such as PET, fMRI, EEG, and TMS, which measure (or excite) activity in very large populations of neurons. Unfortunately, when you are dealing with activity in large populations, you can't tell what any individual neuron is doing, which means a direct positive ID of a mirror neuron is impossible in a healthy human brain. What we *can* do with these methods, however, is document mirror-like properties within a neural *system*, either a local system in a single brain region containing several million neurons or a larger-scale system involving a network of regions and tens or hundreds of millions of neurons. Over decades this is what scientists have done in many experiments using a variety of techniques and targeting a range of human cognitive neuroscience questions.

I've already alluded to the two earliest studies investigating the human mirror system. The two experiments used different methods, TMS and PET, and were published in 1995 and 1996 respectively, nearly coincident with the publication of the 1996 paper on monkey mirror neurons published in the journal *Brain*. These human TMS and PET experiments were referred to in the 1996 *monkey* paper and were used in that foundational report as evidence for the action understanding interpretation of monkey mirror neuron function. But a closer look at the human studies reveals hidden complications for the story the Parma team was trying to tell.

As we saw earlier, the main finding of the TMS study was that viewing a grasping action causes an increase in excitability in the observer's motor system.[1] Stripping away the jargon, this study found that when participants observed a grasping action, their hand was a little more twitchy. In a control condition in which participants viewed graspable objects, increased motor excitability was not detected; the effect was specific to viewing actions. This finding was claimed as evidence for the existence of a human mirror system because it reveals a functional link between perceiving and executing similar actions; perceiving an action somehow facilitates movements that are "mirrors of" that action. However, two things about the study warrant caution with respect to that interpretation. The authors also reported that motor excitability increased during the observation of nongrasping actions (tracing shapes in the air). Monkey mirror neurons only respond to object-directed actions, such as reaching for a raisin, not to pantomimed or other sorts of gestures. If human mirror neurons are behind the results of the TMS study, they are not of the same ilk as monkey mirror neurons. This isn't necessarily a problem with, or criticism of, the study—after all, monkeys and humans are different. But if the monkey and human mirror systems are demonstrably different, then we have to be careful about drawing conclusions in one species based on data collected in another. The 1996 monkey mirror neuron paper, which ascribed similar functions to monkey and human systems, appears to lack this caution. If the systems behave differently, how can we conclude they are performing the same function or even that the measurements are coming from the same network?

his caution

Our caution might be elevated still more by the lack of a motor facilitation effect during *object* observation. This finding was touted by the authors of the TMS study as a virtue of experiment: it spoke to the specificity of the action-observation effect. But it might hint at a problem. To see why, we must take a closer look at the logic of the experiment.

The TMS study team stimulated primary motor cortex to elicit twitches, or motor-evoked potentials as they are called, in the hand muscles. Even though this group also argued that mirror neurons do not live in primary motor cortex (MI)—the region of cortex that most directly connects to the muscles—they reasoned that mirror neurons in Broca's area should be connected to MI, allowing those neurons to exert a facilitory influence on MI neurons when activated by action observation. So it's not that TMS affected mirror neurons directly. Rather, stimulation was applied to the motor neurons in MI simply to make the hand muscles twitch. Then the size of the twitch was measured while the participants observed actions or not. When it was found that the twitch was larger during action observation, it was assumed that this was because viewing actions activated mirror neurons, which activated MI cells, which made the twitch to the TMS pulse bigger than normal. If this logic is on track, then *anything* that can activate the neurons in MI should lead to a larger twitch.

Critically, the mirror neurons were discovered during investigations of another class of cells that respond both during grasping and during the observation of graspable *objects*. These monkey cells, now called *canonical neurons,* live in the same F5 neighborhood as mirror neurons and are believed to support action selection: choosing the right grip type for grasping a particular object. Canonical neurons, like mirror neurons, are connected to motor cells in primary motor cortex. So, following the logic of the TMS study, viewing graspable objects should activate MI neurons just like viewing actions appears to activate MI neurons. But the authors of the TMS study reported that viewing such objects did not result in an increase in twitchiness. Why?

Maybe it is because humans don't have a "canonical neuron system." Unlikely. First, humans grasp objects all day long and there must be a brain mechanism linking object shape to the shape of the hand in a grasping action. Further, experiments have shown that viewing objects of different sizes automatically facilitates related grip types. In one clever study, participants were asked to categorize photographs of objects as either natural or man-made and indicated their response by grasping either a small plastic square requiring a precision grip or a larger plastic cylinder requiring a whole-hand grip. But sometimes the pictured objects and the response grip-type matched and sometimes they didn't. It was found that participants' responses were faster when the response grip-type (e.g., precision grip) matched the size of the pictured object (e.g., a key).[2] Just seeing an object seems to activate an appropriate grip for grasping it.

If humans have a "canonical neuron system," then viewing objects in the TMS study should also have led to a motor facilitation effect just as it did for viewing actions; it did not. Why is this cause for a raised eyebrow? Because it reveals a double standard operating in the interpretation of the TMS results:

> *Actions*: Viewing actions results in motor facilitation as measured using TMS. Therefore, a system must exist in humans that associates observed actions and executed actions, that is, the human mirror system.

> *Objects*: Viewing objects does not result in motor facilitation as measured using TMS. Therefore, humans do not have a system for associating observed objects and executed actions; that is, humans don't have a "canonical neuron system" and shouldn't be able to grasp objects.

In the case of objects, we know that the conclusion can't be right. The finding certainly wasn't interpreted in this way. But the same logic that for obvious reasons wasn't deployed for the object case *is* used to conclude that perceiving actions activates the motor system.

So the TMS study has some complications. It still shows that associations exist between perceiving and executing actions, but we knew that already. What do you do when you meet someone and she thrusts out a hand in greeting, then gives you a business card, and then waves at you when you leave? You shake, grab, and wave back, of course. As the Parma team pointed out in their first 1992 publication, the actions of others are relevant to our own actions and quite obviously the human brain has a mechanism for linking the two. But the existence of an association between perceived and executed actions as in the meet-and-greet scenario above, which is really all the TMS study showed, does not shed light on the real question: Are mirror neurons (or a human mirror system) at the core of our ability to *understand* actions?

The other pivotal early experiment on the human mirror system was a PET study that reported activity in Broca's area during action observation.[3] Broca's area, recall, is the presumed human homologue to monkey area F5. If this region were found to be active during action observation, the logic goes, then we would have evidence for a human mirror system. This is exactly what Rizzolatti and colleagues reported in their first functional imaging study on the *human* mirror system published in 1996.[4] In this study, seven healthy twenty-something-year-old male participants either viewed grasping actions performed live by an experimenter or grasped objects themselves. Brain activation associated with these two conditions was contrasted with a baseline of simply viewing objects held by an experimenter. Consistent with the monkey mirror neuron data, observation of grasping actions resulted in brain activity in two regions, a sensory-related region in the superior temporal lobe and a portion of Broca's area in the left hemisphere. The superior temporal lobe activation was interesting because a similar area in the monkey was found in the mid-1980s—before mirror neurons were discovered—to contain cells that respond during the observation of hand actions.[5] The Broca's area activation got all the glory, however. As in the monkey study, here was a motor-related area that activated during the observation of grasping actions. This seemed like fairly direct evidence for a mirror system in humans.

There was one problem. The Broca's area sector that responded during action observation did *not* also respond during the execution of grasping actions. In fact, there was *no* region that activated both during action observation and action execution, the defining feature of mirror neurons. Instead, when subjects grasped objects (compared to the viewing-object baseline) more traditional motor-control areas were activated: primary motor cortex, adjacent somatosensory cortex (an expected result because subjects were touching the objects), parts of the basal ganglia and the cerebellum (two structures involved in motor control, see Appendix A), as well as a few other regions. While it was no surprise to find the core of the motor system activated during the grasping condition, the lack of activation in Broca's area was not predicted from the monkey mirror neuron data.

In general, the complete lack of overlap between action observation and execution would appear to offer fairly direct evidence *against* the existence of a human mirror system. But that is not how Rizzolatti and colleagues interpreted the data. They instead argued that the simplicity of the task lay behind the failure to find Broca's area activation during action execution: grasping objects wasn't demanding enough to drive activation. They cited several other published experiments on brain activation during action execution, none of which reported activity in Broca's area. But they also noted other studies that *did* report activation in Broca's area during the performance of complex motor tasks such as tapping out a *sequence* of movements rather than a simple grasping action. Despite the lack of a mirror-response pattern in their data, the authors concluded, "The presence of an activation of Broca's region during gesture observation . . . [suggests] a possible common evolutionary mechanism for speech and gesture recognition" (p. 251); that is, mirror neurons exist in humans, they form the neurophysiological basis of action recognition (including speech), and are the missing link that explains the evolution of speech from action recognition mechanisms present in the monkey.

It might seem odd for a study that from one vantage point failed to demonstrate a mirror neuron–like effect to be held up as evidence for the existence of a mirror system in humans. It's not quite so strange. Scientific data are noisy. Our measurements sometimes fail to detect phenomena that really exist and sometimes provide false evidence for phenomena that don't exist. Such outcomes can happen for any number of statistical, technological, or experimental design reasons and it happens all the time in human functional brain imaging studies. The upshot is that we researchers can't interpret every little piece of information from a single study as the final judgment on an issue. Given the relatively small sample size and lack of sensitivity of functional imaging in the mid-1990s, I think it was reasonable for Rizzolatti and colleagues to suggest that Broca's area is involved in executing actions but simply wasn't detected in the 1996 PET experiment. It was a theoretical speed bump and like with the real thing, you can either slow down or hit it at full speed. But already in that 1996 PET experiment there was another speed bump: the location of the Broca's area activation during action observation.

Broca's area comprises two major sectors, a posterior portion called the *pars opercularis* and an anterior zone called the *pars triangularis* (see figure). These two regions are distinguished on the basis of both gross anatomy and internal cellular organization or *cytoarchitectonics* ("cell architecture"). Rizzolatti and colleagues' human PET scan activation during action observation was found in the pars triangularis. This matters because the likely homologue to monkey area F5 is not the pars triangularis but the pars opercularis. So not only did the experiment fail to document a mirror effect—activation for both observation and execution—but the one effect they were able to detect wasn't in the right spot.

Again, this is arguably just a speed bump. The two sectors of Broca's area are right next to each other and functional imaging can sometimes mislocate an activation due to averaging across different brains (see Appendix B). Or, as the authors of the report speculated, maybe the difference reflects evolutionary divergence between humans and macaque monkeys. To compound matters, a companion human subject PET study published by some of the same authors[6]—this one comparing grasping observation with *imagining* the execution

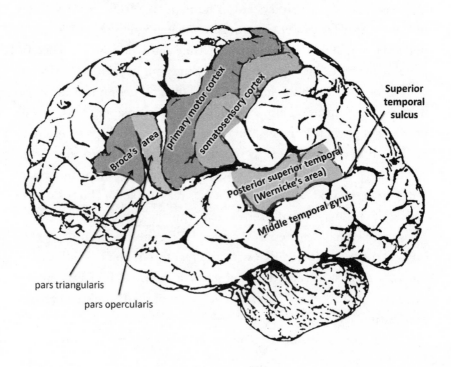

of grasping actions instead of actual execution—reported the same two speed bumps: observing actions activated the pars triangularis, and no overlap between action observation and (imagined) execution. The anomalies replicated. Despite this, the interpretation of the function of mirror neurons and the generalization of the system's existence and properties to humans gained speed in the scientific literature.

To summarize, in 1995 and 1996 four mirror neuron–related papers were published by the Parma team and a few collaborators: the 1995 human TMS paper; the 1996 foundational monkey paper, published in *Brain,* in which the term "mirror neuron" was coined; and the twin 1996 human functional imaging PET papers. These papers laid the foundation for all that was to follow. Although the data didn't line up all that well between the monkey and the human findings, the anomalies did not lead anyone to rethink the functional role of monkey mirror neurons and their relation to a possible similar system in humans. Instead the anomalies were ascribed to complications associated with the tasks or evolutionary differences between humans and monkeys. The conclusion: both monkeys and humans have mirror neurons/systems and they are the neural basis for action understanding. This conclusion, in turn, suggested that mirror neurons could shed new light on complex human abilities.

Let me be clear: there is nothing wrong with a strong, interesting, and creative hypothesis. And there is nothing wrong with pursuing that hypothesis even in the face of a few empirical anomalies, *as long as the research program proceeds with extreme caution.* This was not the case with mirror neuron research. Looking back, what is particularly troubling in those early, foundational publications was the circularity of the arguments:

- Monkey mirror neurons respond both during action execution and action observation. There is more than one possible interpretation, but, argues the 1996 *monkey* mirror neuron paper, humans have a mirror system too, as the TMS and PET studies show. In humans there is a theory in which motor areas are critical for speech recognition. Therefore, monkey mirror neurons must be the basis of action understanding.

- Humans don't show neuronal overlap between action observation and action execution like monkeys do, nor do they show activation during action observation in exactly the right place in the brain, but, according to the 1996 *human* mirror system PET study, given that the "monkey data on F5 neurons suggest that a motor action is recognized by matching it with a similar action motorically coded in the same neuron [the] presence of an activation of Broca's region during gesture observation links the [human and monkey] sets of observations, suggesting a possible common evolutionary mechanism for speech and gesture recognition."[7]

In subsequent publications—including the 1998 game-changing theoretical papers—the troubling details tended to be smoothed over. For example, in both Gallese and Goldman's 1998 paper on mind reading and Rizzolatti and Arbib's 1998 paper on the mirror neuron basis of language, the 1996 PET studies were summarized and cited as evidence for the existence of a mirror system in humans. However, the fact (or more accurately, problem) that there was no overlap between observation and execution in those studies was not discussed. Rizzolatti and Arbib did mention several *other* papers that found activation in Broca's area during the execution of various actions, but these studies did not investigate action observation at all and reported activity mostly in the more posterior sector of Broca's area, the pars opercularis.

While the complications from the early PET studies seemed to be dropped in subsequent theoretical discussions, they remained "rather disappointing," as Rizzolatti and colleagues described them later.[8] True, if you looked *across* many studies you could find evidence for Broca's area activating during observation in one study and during execution in another, but if you were to plot the precise locations of these activations, as a 2001 study did,[9] very little direct overlap was evident in the human homologue of monkey area F5, the core of the mirror system. Although some overlap was evident in other brain regions, such as the parietal lobe, where mirror neurons were subsequently identified in monkeys.[10]

Some evidence for direct overlap in Broca's area finally emerged from a functional imaging study reported in 1999 by Marco Iacoboni and colleagues at UCLA in which human participants viewed and then imitated meaningless hand gestures,[11] but this introduced a further complication to the story. Mirror neurons in macaques do not respond to meaningless actions, nor do macaques imitate the way humans do. If this study was evidence for a human "mirror system," it indicated, again, that in humans the system was doing something very different from monkeys.

THE SEARCH FOR MIRROR NEURONS IN HUMANS CONTINUES

THE LACK of direct evidence for a human mirror system attracted increasing attention in the next decade. Some research teams began to question the existence of mirror neurons in humans, noting the lack of neural overlap between observation and execution and calling out previous work for employing circular logic. In 2008 a collaborative team at NYU and Carnegie Mellon, for example, concluded their review of the state of human mirror system research by writing, "We hope that future human mirror system studies use similar and novel protocols for assessing movement selective responses rather than relying on the circular reasoning commonly used to interpret imitation and passive movement observation experiment results."[12]

The major critique of previous studies that inferred observation-execution overlap was that their functional imaging protocols did not isolate the neural responses they claimed to be isolating. Most functional imaging protocols can only measure the behavior of very large neuronal populations, which means you don't know whether the same *individual* cells that respond during observation also respond during execution.

To illustrate, suppose our target brain region is a city intersection, and our neurons are a large crowd of people standing around the intersection. Being neuron-people they spend their whole day at the same intersection traversing the crosswalks ("firing") whenever their light turns green. We want to measure the activity of the people-neurons but can only do so via a device that detects the sum of the motion within the entire intersection. We notice that when one crossing light turns green, say the north–south light, we detect a strong motion signal. And when the other, east–west light turns green we also detect a strong signal. Our intersection of people-neurons "activates" both during north–south "stimulation" and during east–west "stimulation." Are the same people-neurons responding to both types of stimulation? You can't tell. It could be that all of the people-neurons cross back and forth only within their preferred crosswalk, or everyone could be crossing in both orientations, "activating" both to north–south and east–west stimulation. The same is true for standard fMRI measurements: even if you find a region that responds to two different conditions, you can't tell whether the same neurons are responding to both, or whether there are two intermixed populations of neurons.

There's a way around this limitation. Neurons adapt or habituate to repeated stimulation—the neural basis of "the water feels good once you get used to it!" If, in our intersection analogy, our people-neurons habituate we should expect that those people who have to cross repeatedly without rest will do so less vigorously than those who get a break between crossings. We could then conduct the following experiment. Turn on the north–south crossing light two times in a row and measure the difference in the people-neuron response to the first versus the second light; the second response should be less vigorous, showing habituation. Now turn on the north–south light followed by the east–west light. If no habituation is observed, we can infer that the people-neurons crossing in the two orientations are not the same: the crossers are fresh, not habituated, and respond equally vigorously to their signal. If habituation *is* observed in our measurement, we can conclude that at least some of the people-neurons are crossing in both orientations because *someone* got tired having to cross twice in a row. We still can't tell who exactly is going where, or how many are doing what, but at least we know that a subpopulation of individual people-neurons is going both ways.

What happens when you apply this *fMRI adaptation* method, as it is called, to the question of mirror neurons? Do any regions show habituation when action observation and action execution are paired? The first couple of attempts to answer this question suggested *no*, and this fanned the embers of doubt regarding the role of mirror neurons in human behavior.[13] But these studies involved the imitation of hand gestures, not the object-directed actions that monkey mirror neurons prefer. Finally, in the summer of 2009, more than a decade after the first bold speculations about the existence of mirror neurons in humans, a team of researchers at University College London published a report that provided the first relatively direct evidence for the existence of mirror neurons in the presumed human homologue of monkey area F5. In 9 out of 10 individuals who were studied, the posterior sector of Broca's area exhibited habituation effects across the observation and execution of object-directed actions.[14]

Not long after this 2009 report, a team at UCLA reported the first and so far only direct demonstration of mirror neurons in humans. Twenty-one patients undergoing surgical treatment for pharmacologically intractable epilepsy were studied. As part of the clinical procedure to identify the source of the seizures, electrodes were implanted in the patients' brains, which allowed recordings from individual neurons. Video clips of grasping actions were shown to the participants and they were also asked to perform grasping actions themselves; a similar facial gesture observation-execution task was also used. The UCLA team reported that individual cells were found to respond both during action observation and execution in several brain regions located near the midline of the two hemispheres, including one motor area, known as the supplementary motor area or SMA. The monkey homologues of these regions have not been assessed for the presence of mirror neurons, but at least now there is direct evidence that some cells, somewhere in the human brain, respond in a mirror-like fashion.

But this finding is not all that helpful for trying to unravel *what* mirror neurons are actually doing in macaque monkeys and humans. To answer that question, we must dig deeper into the idea of motor simulation as the basis for action understanding. And once you start digging, you find some troubling anomalies.

4

Anomalies

TURN YOUR attention to the sky and it looks like the Sun revolves around Earth. This evidence for an Earth-centered view of the heavens is so compelling that it is no surprise that the geocentric theory dominated philosophy, science, and religion for centuries. Over time astronomers noticed anomalies that didn't fit the theory. The major anomaly was retrograde planetary motion. Objects in the night sky, including both stars and planets, move from east to west over the course of a night. However, planets observed from one night to the next tend to move from west to east *relative to the backdrop of stars*. This action is where the term *planet* comes from, part of a Greek phrase meaning "wandering star." Occasionally, planets reverse direction for a few weeks relative to the stars. This anomalous *retrograde motion* makes no sense in a model of the heavens in which all the celestial bodies revolve around Earth.

Ptolemy proposed a convoluted but geocentric-preserving solution in the second century that involved planetary orbits organized around a series of concentric spheres instead of tidy circular orbits, but it wasn't until the sixteenth century that retrograde planetary motion finally spurred Copernicus's heliocentric revolution. By placing the Sun at the center of the planetary orbits, Copernicus showed that satisfyingly simple planetary orbits were consistent with retrograde planetary motion because the retrograde motion is only *apparent*, not real. Planets look like they reverse their orbit because we are observing those orbits from a moving platform. When Earth's orbit overtakes that of another planet, like passing a car on a highway, the planet appears to move backward.

What does this have to do with mirror neurons? It illustrates the substantial pull of a commonsense intuition. Because the geocentric theory is so intuitively consistent with the in-your-face fact of the Sun "moving" across the sky, it took centuries to come to understand what is really going on. We may be in a similar situation with mirror neurons. Mirror neurons fire *both* when the monkey executes an action *and* when it observes a similar action. Given this glaring fact, mirror neurons *must* be involved *both* in generating movements *and* in understanding the actions of others. But there are anomalies. A few of them were highlighted in the previous chapter: the human and monkey mirror systems appear to behave differently, which complicates cross-species inferences. Once you start looking closely at the evidence, you see plenty more.

ANOMALY I: SPEECH PERCEPTION WITHOUT THE MOTOR SPEECH SYSTEM

THE MOTOR theory of speech perception, developed in the 1950s, was a major inspiration for the mirror neuron theory of action understanding (details of this backstory appear in Chapter 5). In a nutshell, the motor theory asserts that we perceive speech not by recognizing the sounds, as one might think, but by recognizing the vocal tract motor gestures that shaped those sounds. From the very first reports of the monkey experiments in 1992 and 1996, mirror neurons were touted as a possible neurophysiological mechanism for the motor theory of speech perception, making language the first aspect of human behavior that mirror neurons supposedly explained. The problem with the motor theory of speech perception and its mirror neuron instantiation is a well-known type of language disorder called Broca's aphasia.

Broca's aphasia is characterized by impaired ability to produce speech after damage to motor speech centers in the brain including Broca's area, the claimed homologue of macaque area F5. If the motor theory/ mirror neuron claims were correct, namely that motor speech areas are critically involved in speech perception, we would expect that patients with Broca's aphasia would have speech recognition problems that parallel their speech production deficits. Yet, they do not. People with Broca's aphasia have quite good auditory comprehension of speech. In fact, Broca's original case report involved a patient referred to as "tan," the only syllable he could speak. Yet Broca remarked that his now famous patient could understand virtually everything said to him. If the mirror neuron theory of action understanding were correct, Broca's aphasia should not exist.

ANOMALY 2: UNDERSTANDING ACTIONS THAT WE CAN'T PERFORM

I HAVE two cats and a dog and here's what I know. When my dog wags his tail, he is happy. When he tucks it between his legs, he is scared. Perky ears mean attentive, ready, but when they are pinned back, something is wrong. He has several barks and other vocalizations that mean aggression, excitement, "that's mine don't touch," sadness, and fear. When my cat purrs she is socially receptive; when she hisses, I need to back off. A cat tail whipped from side to side indicates annoyance. A tail held high and vibrated means socially receptive. How do I know what these actions mean? I don't bark or purr, I can't move my ears, and I don't have a tail.

That we can understand the actions of our pets, the flight of a bird, the coiling or rattling of a snake, the jet propulsion of a squid, or the reverse slam dunk of an NBA star tells us that we don't need to be able to perform an action to understand it. In fact, from an evolutionary standpoint, we had better be able to understand and predict the actions of our predators and prey even if they don't move like us lest we get eaten or go hungry.

Mirror neuron theorists acknowledge this fact. In an influential 2004 review article[1] on the mirror neuron system, Rizzolatti and Craighero admit that the brain can understand actions without the mirror system; "although we are fully convinced . . . that the mirror neuron mechanism is a mechanism of great evolutionary importance through which primates understand actions done by their conspecifics, we cannot claim that this is the only mechanism through which actions done by others may be understood" (p. 172). This raises a problem, though. If another mechanism exists, what selection pressure could have led to the evolution of mirror neurons? If the cells are effectively redundant, then there is nothing for natural selection to work with. They have to have conferred survival benefit beyond that of the "other" action understanding network.

A speculation regarding the extra benefit has been proposed as "understanding from the inside," the idea that we don't truly understand an action unless we can connect with it motorically. So we can recognize flying, but we don't *really* understand it because we've never flown under our own power. Here's an explanation of the claim from a 2010 review paper by Rizzolatti and Corrado Sinigaglia.[2] It follows the authors' summary of a previous fMRI study[3] showing that observers can recognize animal actions outside the human motor repertoire (e.g., barking) and that such recognition involves the *ventral stream* (see Appendix A), part of the temporal lobes, and not the mirror system:

> These data indicate that the recognition of the motor behaviour of others can rely on the mere processing of its visual aspects. This processing is similar to that performed by the "ventral stream" areas for the recognition of inanimate objects. It allows the labelling of the observed behaviour, but does not provide the observer with cues that are necessary for a real understanding of the conveyed message (for example, the communicative intent of the barking dog). By contrast, when the observed action impinges on [activates] the motor system through the mirror mechanism, that action is not only visually labelled but also understood, because the motor representation of its goal

is shared by the observer and the agent. In other words, the observed action is understood from the inside as a motor possibility and not just from the outside as a mere visual experience. (p. 270)

Dog owners or casual dog observers may take issue with this claim. It is not all that difficult to discriminate the meaning behind various types of barks, from excitement to aggression. Empirical evidence backs up this intuition. A team of scientists recorded the barks of Hungarian sheepdogs in various situations: when the dogs were approached by a stranger, when they were preparing to fight, when they were left alone, and when they were about to play ball, among others. The group then asked 45 individuals to listen to the barks and attempt to categorize them into the situations in which the recordings were made. Performance was well above chance levels for categorizing the stranger, fight, alone, and play barks. Errors tended to be within similar categories (such as confusing a "stranger" bark with a "fighting" bark). Study participants also rated the emotional quality of the barks. Stranger barks were rated to be more aggressive and least happy and playful, alone barks were characterized by high scores for fear and despair, and play barks were judged as the most happy and playful. What's more, the study included three groups of participants, sighted individuals, blind individuals with previous visual experience, and congenitally blind individuals (therefore no visual experience). The groups performed similarly, an indication that "understanding" the intention behind the bark could be achieved using the acoustics alone and not on the basis of previously experienced visual cues associated with the barks.[4]

From an evolutionary standpoint this result makes sense. If one species can read the various types of actions of another species, either innately or through learning, this provides a substantial survival benefit. The question is, what does "understanding from the inside" add to this ability? According to Rizzolatti and Sinigaglia it tags the action as "a motor possibility," in other words, the realization that *Hey, I know how to do that too!* While this might be beneficial for some motor tasks, such as learning behaviors from observing others (so-called observational learning, a point we return to in Chapter 8), it seems peripheral to understanding actions at a fundamental level, like determining whether a dog is going to attack or lick. Further, it is worth noting that the narrower claim that mirror neurons allow "understanding from the inside" is a rather different and substantially weaker claim compared to the popular idea that mirror neurons are *the* basis for action understanding.

ANOMALY 3: LIMB APRAXIA

LIMB APRAXIA is a neurological disorder, typically caused by stroke or neurodegenerative diseases such as Alzheimer's. It impairs an individual's ability to perform tasks or movements when asked to do so even though the patient understands the command, is willing to perform it, is familiar with the task, and has no muscular problem that would prevent execution of the movements. In short, it is a deficit involving the execution of skilled, purposeful movements. Pantomime and imitation of gestures prove the most difficult, while real-world actions involving actual objects are typically much less affected. A classic example is an apraxic patient who is asked, "Show me how you would answer the phone." He brings his hand toward his head but touches his fingers to his mouth like he's eating something instead of holding an imaginary phone to his ear. He might then fumble around unsuccessfully for the right gesture. When the phone rings, however, he picks it up and answers it without trouble.

The 1992 and 1996 monkey mirror neurons papers invoked limb apraxia as a possible bridge to systems involved in human action understanding. The basic idea: patients with limb apraxia have deficits in performing actions and some of them also have deficits in recognizing actions. This was considered suggestive neuropsychological evidence for the claim that action execution systems are the basis for action understanding. *But,* apraxia mostly affects pantomime and imitation, two abilities that monkeys don't have and mirror neurons do not respond to. Therefore, even if an association could be firmly established between action execution and action understanding in apraxia, it would reflect the function of a system that is different from the monkey mirror neuron system. Nonetheless, to show *some* relation between action control and action understanding would be a substantial step forward for the mirror system program.

Predictably, limb apraxia has become a hot topic in mirror system research. A 2008 study in particular,[5] carried out by a team in Rome, has received a lot of attention. It seems to provide strong evidence for the sought-after causal link between action execution and action understanding. The team assessed 41 patients who had suffered a stroke in the left (33) or right (8) hemisphere on their ability to execute and comprehend gestures. Limb apraxia is primarily associated with left-hemisphere injury so we focus here on the group of 33. In the left-hemisphere group, 21 patients were classified as apraxic after researchers observed deficits either on a gesture imitation task (such as mimicking a hitchhiking gesture) or on a task that involved the performance of a sequence of movements with real objects (e.g., lighting a candle). Consistent with previous research, deficits involving real objects were less severe on average (82 percent accuracy) than deficits on the imitation task (54 percent accuracy). This means that the correlations noted below are most likely driven by performance on the imitation task. The 12 patients without apraxia scored 100 percent on the object task and 91 percent on the imitation task.

Here's the critical question for testing the mirror system hypothesis: Do the patients with limb apraxia have deficits in recognizing gestures? To assess this, the study team showed the patients video clips of an actor performing object-directed actions such as strumming a guitar and non-object-directed actions such as making a hitchhiking gesture. Patients also viewed a set of incorrect-action clips. Incorrect actions showed, for example, the actor strumming a flute instead of a guitar or hitchhiking with the pinky finger extended instead of the thumb. Patients were asked to indicate whether the action was correct or incorrect.

If the mirror neuron theory of action understanding is correct (ignoring for now complications concerning imitation), patients with limb apraxia should have deficits in recognizing actions; moreover, the more severe the apraxia, the more severely impaired their understanding should be. This is precisely what the study reports. Apraxic patients scored 68 percent correct whereas non-apraxic patients scored 90 percent correct on the recognition task. Furthermore, a strong positive correlation was reported between gesture execution and gesture recognition scores: the greater the expressive deficit (more trouble executing actions), the greater the receptive deficit (more trouble categorizing videos as correct versus incorrect actions). However, averages can obscure important details and correlations do not necessarily reflect causation. As mathematician Des MacHale noted, "The average human has one breast and one testicle." To this we might add, left and right breast size is strongly correlated, but this doesn't mean the left one caused the right one to grow.

Indeed, a closer look at this study uncovers some revealing details. One is that 7 out of the 21 patients with apraxia did not have any action recognition deficit; that is, they had trouble executing actions but had no trouble recognizing the difference between correct and incorrect actions in the videos. This represents fully one-third of the apraxic individuals in their sample—a sizable minority that cannot be disregarded as statistical noise. What this shows conclusively is that the ability to understand actions does not require the ability to execute them, contrary to the predictions of the mirror neuron theory of action understanding.

Some questions remain. How do we explain the remaining two-thirds of the sample who show deficits in both execution and understanding? And what about the correlation between the severity of expressive and receptive deficits? There are many possible explanations. Here's one: the critical brain region(s) involved in action understanding are distinct from those involved in motor control, but in some apraxic patients they are both damaged. If the two circuits are in relatively close proximity, deficits may pattern together, leading to significant correlations or average effects that look like they are coming out of damage from the same circuit.

To summarize the findings from apraxia in a sentence: Deficits in the control of action do not uniformly result in deficits in understanding of action and this is an anomaly for the mirror neuron theory of action understanding.

ANOMALY 4: MOEBIUS SYNDROME

WHEN WE are happy, we smile; when angry, we frown; when surprised, well, we look surprised. Emotions trigger specific and easily recognizable facial expressions. At least seven of them—anger, contempt, disgust, fear, happiness, sadness, and surprise—are recognized across all human cultures.[6] Recognizing facial expressions has obvious social implications and is a primary source of information for mind reading, understanding what others are thinking or feeling. This is exactly the kind of ability that the human mirror system is supposed to support, and motor-simulation theorists have a ready explanation for how we read emotions on the faces of others. Because we already have built-in neural associations between our own emotions and the corresponding facial expressions (e.g., happy emotions trigger happy expressions), we can read others' facial emotions simply by simulating the facial expression we observe and running the association in reverse. By simulating the expression, we tap the associated emotion in ourselves and literally feel the pain or joy or surprise of the person we are observing.

There is evidence that this might be true, that the emotion-face relation can work backward. American psychologist Paul Ekman and colleagues found that simply asking people to move their face into certain configurations without mentioning the particular emotion (lower your eyebrows, narrow your eyes, and so on) can induce physiological responses associated with the facial configuration; for instance, an anger-related configuration can trigger an increased heart rate.[7]

But what happens if you don't have those automatic associations between emotions and facial expressions? What if you can't express emotion on your face at all? This is the situation in which individuals with Moebius syndrome find themselves.[8] Moebius syndrome is a rare congenital, bilateral paralysis of the face that makes it impossible to express facial emotion. Its cause is unknown. Inability to move the lips affects an individual's speech articulation as well, but with the rest of the vocal tract functioning, this hurdle can be overcome with speech therapy. Most affected individuals can learn to speak clearly.

It's hard to imagine how difficult social communication would be without facial expressions. As noted by Kathleen Bogart and David Matsumoto in a recent article, people with Moebius syndrome are unable to participate in one of the few universal languages, the expression of facial emotion. A friendly smile or wide-eyed interest is out of the question. How do you communicate what you are feeling? You can say it verbally, but who would believe you when the words are coming out of a slate-blank face? To those who don't know you, you would seem unfriendly and robotic. It's no surprise then that affected individuals tend to be more inhibited, introverted, and report more feelings of inadequacy and inferiority.

The motor effects of Moebius syndrome have been known since the syndrome was first described in 1888 by German neurologist Paul Julius Moebius. But only recently have the perceptual effects of the syndrome been investigated. The question is whether the inability to express facial emotion motorically prevents individuals with Moebius syndrome from recognizing emotions in others. The motor simulation account of action understanding would predict that it does.

In a 2010 study published in the journal *Social Neuroscience*, Bogart and Matsumoto tested 37 adults with Moebius syndrome and 37 matched control subjects on a facial expression recognition task.[9] Participants viewed a set of 42 color photographs of faces expressing various emotions, one at a time; the participant's task was to identify which of the seven universal emotions each face expressed. The data showed that the Moebius syndrome group performed just as well as unaffected control subjects. No statistically significant differences were found. In fact, the Moebius syndrome group actually achieved higher raw scores than the control group on five of the seven emotion categories. In other words, there is no deficit in the recognition of facial emotions among people with Moebius syndrome. In contrast to the predictions of the motor simulation theory, you don't need to be able to express facial emotions to recognize them.

Motor-simulation theorists invoke a counterargument to this kind of result. It goes like this: Moebius syndrome affects the ability to implement a motor plan due to dysfunctional facial nerves, but the motor plan itself lives in the cerebral cortex and is unaffected. Therefore, individuals with Moebius syndrome can still simulate *internally* the facial movements they see, and this higher level is where their recognition abilities reside. This is a legitimate possibility and a potentially viable argument.

There is, however, a counter-counterargument. The cerebral cortex is a highly plastic and adaptable network that can reorganize according to its particular sensorimotor situation. For example, if the inputs to a particular region of cortex get cut off for some reason, that chunk of tissue often starts responding to stimulation from a different source.[10] If you were unlucky enough, for example, to lose your middle finger in a freak road-rage incident, the cortical tissue that used to respond to middle finger sensory stimulation would instead become responsive to inputs from your index and ring fingers, both of which occupy nearby cortical regions. Having lost its normal sensory input, the tissue is effectively reprogrammed to process sensory information coming from nearby regions of the body, as if neighboring cortical territories invade the unused tissue. We know, further, from situations involving congenital and chronic deafferentation, as loss of input is called, that a cortical area can "forget" (or fail to learn) how to process information from its normal source. A classic example is congenital peripheral deafness, hearing loss due to dysfunction of the inner ear (cochlea) or auditory nerve. With the development of cochlear implant technology, peripheral auditory stimulation can be restored to a level that can allow for reasonably good hearing and even support speech ability. However, the age at which the device is implanted matters. One study reported abnormal auditory responses in children who received implants after age seven[11] and there is evidence that children who receive implants before two years of age have greater benefit than those implanted between ages two and three.[12] Congenitally deaf individuals who receive their first implant as adults fare substantially worse: their ability to recognize speech improves minimally if at all.[13]

The point here is a familiar one—use it or lose it—and it raises the question whether individuals with Moebius syndrome, having never used it, might in fact have lost it, that is, the cortical system controlling facial expression. If true, this possibility would undermine the motor simulation counter argument about a motor "plan" that resides in the cortex.

ANOMALY 5: MIRRORING CAN BE MALADAPTIVE

MIRROR NEURON theorists tout the virtue of "direct" understanding via motor simulation *without the need for complex cognitive inference.* For example, if you observe a friend reaching toward a raisin, there is no need to run through the logical possibilities—*Ally likes raisins, Ally is hungry, Ally is probably reaching for the raisin to eat it*; by simulating her movement, you understand it "directly" and "automatically." This might be considered a theoretical plus since explaining complex things using simple mechanisms is, after all, a good thing, but it also has potentially catastrophic consequences. If we had to simulate the actions of others in our own motor system in order to understand them, this could interfere with nonmirror actions that we may need to respond with.

Sports provide a good example. In many sports, athletes must respond to the actions of other players, typically in a nonmirrored fashion. A batter in a baseball game must prepare a swing in response to observing a pitch. A boxer must duck or block an incoming punch. A goalkeeper must lunge or jump upon seeing a kick. Examples like this abound and in most of them, reaction time is the key to success. Now, if the batter, the boxer, and the goalkeeper had to first simulate the actions he or she observed, this would tend to slow reaction times because the motor system would have to activate a simulated program that competes and interferes with the one required to successfully respond to the action. We saw an experimental example of such an interference effect in Chapter 3 (the study in which subjects made judgments about large or small objects and had to respond with either larger or smaller grips) and we'll see another in Chapter 6. Mirroring is therefore potentially maladaptive in situations that require fast, nonmirror reactions in response to observed actions.

ANOMALY 6: THE MIRROR SYSTEM IS MALLEABLE

IN 2007, a team of scientists at the University College London—Caroline Catmur, Vincent Walsh, and Cecilia Heyes—carried out the following TMS experiment.[14] They first replicated the basic TMS mirror neuron result by showing that a subject's index finger is more twitchy while watching videos of index fingers moving than while watching videos of pinky fingers moving (and vice versa). The team then set about training their participants to make "reverse mirror movements" in response to the videos of index or pinky finger movement: when subjects observed an index finger movement, their task was to move their pinky finger, and vice versa, over 864 trials. Then, again using the standard TMS experimental paradigm, Catmur et al. measured the subject's brain response to viewing finger movement and found that the mirror effect of observation had been reversed. Viewing *index* finger movement made the observer's *pinky* more twitchy and viewing *pinky* movement made the *index* finger more twitchy, even though subjects were asked to simply view the movements and *not* perform a movement task themselves.

Just like a feigned punch to the face generates an anticipatory flinch on the part of the punchee, participants in the TMS study generated their own (covert) anticipatory motor response reflecting the observation-execution pairing (index/pinky, pinky/index) that they were trained on. It's simple stimulus-response learning. A red light dangling from a post in the middle of an intersection doesn't in itself require that you lift your foot off one automobile pedal and push down on another, but after years of driving experience, your brain has established an association between light color and leg movements. If we used TMS to measure leg muscle responses to viewing red lights, it would no doubt show an increase in leg twitchiness when viewing a traffic light that changes color, even when your task is to sit quietly in the passenger seat. Catmur, Walsh, and Heyes basically demonstrated that the "mirror system" is capable of the same type of stimulus-response associative learning.

Why is this a problem for the mirror neuron theory of action understanding? Because it's probably a safe bet that when the reverse-mirror-movement-trained subjects viewed a pinky movement (and gained a twitchier index finger as a result), they didn't "misunderstand" it as an index finger movement. The sensorimotor training did not alter participants' perception, it only altered the *motor response* triggered by the action observation, the sensorimotor pairing. Action execution and action understanding dissociate: if you can alter the motor response without affecting perception, then the motor response can't be the basis of the percept.

Jonathan Venezia, a graduate student in my own lab, took the concept of remapping the "mirror" stimulus pattern to the extreme. Reasoning that TMS-induced "mirror responses" reflect nothing more than associative pairing, he conducted an experiment similar to the London group's. First, he showed the standard "mirror effect": viewing pinky movement causes an increase in pinky twitchiness and viewing index finger movement causes an increase in index finger twitchiness. Then, instead of training subjects to make reverse mirror movements, he trained them to make pinky movements in response to viewing a picture of a cloud and make index finger movements in response to viewing a picture of a building (or vice versa). Question: does the "mirror system," as measured by TMS, even require action observation? Or does it respond to any old sensory stimulus that is paired with an action? Will cloud viewing lead to pinky twitches and building viewing lead to index finger twitches? Answer: yes. The human "mirror system" is so malleable that it doesn't even need to involve action observation. You can pair an arbitrary stimulus, even one that you can't grasp, with a particular hand movement and after only an hour or so of training, simply observing the stimulus primes the associated motor response.

In light of these findings, let's revisit the logic of the mirror neuron theory of action understanding. The mirror system activates both during action execution and action observation. According to the theory, this activation pattern reflects a mechanism in which others' actions are understood by simulating those actions in our own motor system. But the "mirror system" also activates both during action execution and cloud observation. Does this mean we understand clouds by simulating the perception of clouds in our own motor system? Obviously not. So if it doesn't work for clouds, we have to wonder whether it works for actions.

There's another possible conclusion, though. Maybe TMS isn't accurately measuring the function of the mirror system, or is measuring different subcircuits, one being the "real" mirror system that codes actions and another that can code arbitrary mappings between sensory events and actions. This is a serious possibility, but if true, it raises the question of what exactly is being measured in *any* TMS study of sensorimotor interactions.

Happily, TMS is not the only source of evidence for mirror system plasticity. The malleability of the mirror system is evident also in single cell recordings in monkeys. Recall the foundational 1996 study on monkey mirror neurons, which reported that the cells respond to the observation of grasping actions made by human hands but not when the grasping was performed by a tool, such as a pair of pliers. This was taken as evidence that the system is specific to actions that are in the monkey's repertoire—monkeys don't use pliers—thus supporting a mechanism that matches or simulates observed actions with executable actions. In 2005, the assertion was revised based on another experiment published by the Parma group:

> Observations made in our laboratory showed that at the end of a relatively long period of experiments, it was possible to find mirror neurons responding also to actions made by the experimenter with tools.[15]

In this experiment, as in the original experiments, monkeys were trained for two months on tasks involving the observation of objects and actions and the execution of actions directed toward the objects. However, during the training for this experiment, the monkeys also observed actions involving two tools: a stick that could be used for poking objects and pliers used for grasping objects. These tools were used to pick up food and give it to the monkey. The authors report that the stick tool was used more often during training.

In the main experiment, recordings were made from 209 neurons in the F5 region of two monkeys during action observation (with tools and with just hands) and action execution. The basic mirror neuron result was replicated: most of the neurons responded during action execution (greater than 95 percent) and a fraction of these responded also during action observation. Seventy-four of them behaved like typical mirror neurons, responding best during the observation of nontool actions. However, 42 (20.1 percent) were found to respond best to the observation of tool actions and were thus dubbed *tool-responding mirror neurons*. Thirty-three of these tool-responding mirror neurons were studied further. It was noted that the majority, 26, preferred the observation of stick actions, while 4 liked pliers better and 3 went both ways. Twenty-nine of the 33 cells also responded to the observation of regular, nontool actions, although more weakly.

It is unlikely given the high percentage of the tool-responding mirror neurons in the total sample that previous experiments simply missed these cells. The two months of training must have driven their existence, which is to say before training they did not respond to the observation of tool actions and only acquired the response to tool actions after prolonged training. The fact that there were more stick-preferring mirror neurons supports this conclusion, given that more training occurred with stick actions. The authors also report that the tool mirror neurons were not detected until the latter stages of the experiment, that is, after even more training occurred.

What this finding means is debatable. The authors of the report argue that tool-responding mirror neurons reflect the system's ability to generalize the understanding of actions from, say, hand grasping to tool grasping. They suggest that the neurons have acquired the ability to treat the tool as an extension of the hand and therefore underpin a new ability in the monkey to understand tool-assisted actions. It's a possible interpretation. However, it is not at all clear that the monkeys really "understood" the tool use at the level implied by the study's authors.

After the recording sessions were done for one of the monkeys, the research team positioned a platform in front of the monkey's cage. Morsels of food were placed on the platform, just out of reach of the monkey. The researchers waited 10 minutes and let the monkey make some unsuccessful attempts to reach the food. Then, they placed the stick used in the experiment on the platform, within the monkey's reach. They left the room and watched the monkey for the next hour via a video monitor. The authors report, "The monkey never attempted to use the tool for reaching food, although in the first minutes after the stick was available, the monkey grasped it and bit it." Instead, the monkey ignored the stick and tried to get to the food by moving the platform. So tool-responding neurons were trained into the monkey's brain and unambiguously recorded. But the monkey's post-training behavior provides no indication that the monkey understood that the tool could be used to poke and acquire food. The mirroring behavior of the tool-responding mirror neurons did not translate into "understanding," at least not an understanding that led to action.

This tool experiment reveals a potentially critical insight that may help us understand where regular mirror neurons come from and what they are doing. The original discovery of mirror neurons was made after hours of training on a reaching task during which the monkeys observed an experimenter using his hand to place and grasp objects of different sizes and shapes. If the observation of hand actions (like tool actions) becomes associated with the execution of a monkey's own actions—that is, if a monkey observes hand actions that are consistently relevant to the monkey's own actions (*when the hand places an object, I grab the object*)—then a cell that lives in a sensorimotor association circuit for the purposes of controlling action may start to respond to the observation of the action alone. And it does this only when there is an object involved (because that is what is relevant for the monkey's own actions). We discuss this possibility in greater depth in Chapter 8.

ANOMALY 7: MIRROR NEURONS ARE FUNCTIONAL OUTLIERS IN THE ORGANIZATION OF CORTICAL SYSTEMS

TO SEE why mirror neurons don't fit within the functional organization of the cerebral cortex we need to traverse a brief digression concerning our current understanding of what the functional organization of cortex looks like. Before we jump in, let me explain what I mean by "functional" in this context. The distinction between neuroanatomy, the study of brain structure in its static state, and neurophysiology, the study of the brain in action, is well known. Sometimes the term *functional* is used quasi-synonymously with physiology. This is the case with *functional* MRI, which measures a part of the brain's physiology, namely blood flow patterns, in contrast to *anatomical* or *structural* MRI.

But brains are more than just anatomy and physiology; they are designed to perform a range of tasks, or *functions*, like controlling respiration, sleep-wake cycles, perceiving objects visually or auditorily, moving, talking, and remembering, among many others. It is often assumed that the brain accomplishes various functions by performing *computations*, much like computers accomplish particular functions by performing computations (more on this point in Chapter 6). When cognitive neuroscientists talk of *functional organization* or *functional systems* they are referring to the different tasks that the brain performs and how the systems or networks or areas that support these tasks are organized within the brain. Thus, we can talk about *functional anatomy*, which refers to the relation between various functions and anatomical structures. For example, that the auditory system occupies the superior temporal lobe whereas the visual system inhabits the occipital lobe (and beyond) is a statement about functional anatomy. Our focus here is on the functional organization within the cerebral cortex.

In the last couple of centuries, we've learned that the functional organization of mammalian cortex is not a collection of circumscribed areas each responsible for its own separable function. This is the myth of phrenology, which held that cortex was composed of dozens of functional islands, mini neuroorgans, each one dedicated uniquely to a specific ability or trait. While phrenology was right about the cortex being nonuniform in function, it was wrong about the neuroarchipelago architecture. Instead, systems that support a given function are organized into distributed networks involving dozens of interconnected subregions that communicate reciprocally (not in a unidirectional flow of information) and often interact with other functional networks (e.g., language is useless without memory). Another thing we've learned is that these functional networks are not plopped down in the brain willy-nilly. Rather they are organized into broad processing streams, partially segregated information superhighways that share functional properties.

You probably are familiar with one principle of cortical organization: that the left brain is more logical, verbal, and detail oriented while the right brain is more creative and passionate, spatial and holistic. This is also a bit of a popular myth, the twentieth century's equivalent of phrenology. It is based on the observation that the two hemispheres of the brain are not perfectly symmetric in function. In pop culture, this perfectly valid observation has been blown out of proportion, leading to the idea that the left hemisphere is the neural version of a logical, fun-sucking math nerd while the right hemisphere is the creative, fun-loving artist. In fact, the two hemispheres are much more functionally similar to one another than are different circuits *within* a given hemisphere. For example, both hemispheres have auditory and visual systems. On close examination, one might be able to tease apart some subtle differences between the left and right sensory systems; some evidence suggests that the left hemisphere may be somewhat better at processing the details of the sensory environment, while the right may emphasize the big picture. But they are subtle indeed compared to the anatomical and functional differences between the auditory and visual systems *within* a given hemisphere.

Beyond the obvious within-hemisphere differences among sensory systems, other functional anatomic differences cut across systems. A major organizational principle is that between the so-called ventral "what" versus dorsal "how" processing streams.[16]

The brain needs to do two basic kinds of things with sensory information. One is to understand *what* it is sensing and the other is to know *how* to act based on what is being sensed. These constitute different functional goals that involve distinct neural circuits. Consider the different tasks a brain faces when presented with the visual image of a cup. To recognize *what* the object is, the visual system has to ignore information about the cup's size, the details of the object's shape, its location in space relative to other objects and the body, and its orientation. After all, cups come in many shapes and sizes and they don't stop being cups just because they are near or far, behind or in front, upright or sideways. The visual system needs to extract general features—learned on the basis of previous experience with cups—to identify an object as a cup. That's the *what* function. Often we need not only to recognize a cup as a cup but also to grasp it. Suddenly the details that didn't matter for recognition matter hugely for grasping. We need to know precisely how big the object is, what its particular shape is, whether there is something in front of it that obstructs our reach, where it is in space, and whether it is upright or on its side. That's the *how* function.

Research in both visual and auditory domains has demonstrated that the neural circuits that support the *what* and *how* functions are organized into anatomically distinguishable networks.[17] In the visual domain, damage to ventral brain regions including the occipital and temporal lobes can lead to *visual agnosia*, a deficit in the ability to recognize objects by sight.[18] It's not that agnosia subjects can't see the objects—it's that they can't recognize *what* they are. Patients may be able to identify general shapes or features, but the essence of the object escapes them. They sometimes describe the objects as appearing "nondescript." A physician with visual agnosia, for example, once described a glove as a container with five outpouchings and described a stethoscope as a long cord with a disk at one end. He had no idea what the objects were, however. The deficit also is not general but specific to visual recognition. If you show a visual agnosia patient a set of keys he may have no idea what they are, but as soon as he picks them up, he immediately recognizes them. What's interesting is that the ability to grasp objects appropriately, to know *how* to interact with them, can remain unaffected.[19]

Damage to more dorsal regions of cortex in the parietal lobes can produce a nearly complementary pattern of symptoms. In a disorder referred to as *optic ataxia*, patients can recognize objects with ease, but reaching for them with visual guidance can prove rather difficult. In severe cases, patients may grope as if in the dark, in the attempt to latch on to their intended target. The deficit is not absolute. It emerges most often when affected patients reach for objects in the visual periphery, or when changing conditions require rapid adjustments, or when the object is unfamiliar[20]—precisely those conditions when the particulars of the object, as opposed to the *what* of the object, are critical for guiding action.

If I asked you to show me how you reach for a cup, you could do it in a fairly generic way. Most likely, you would reach straight out in front of you, at about the height of a table that you are hypothetically sitting in front of, with your hand oriented for an upright cup with an aperture roughly the size of a Starbucks grande. Essentially, you have reached for so many cups in your life that you have a stored motor program for how to reach for the average cup in its average location. Once such a program is acquired, you don't need visual input from the cup to do this. Now, if we put you in a real reaching situation, and I place an object in your periphery and ask you to reach without looking directly at it, you have to modify your default reach and for that you need visual input. Likewise you have to deviate from the default mode, using visual details, if you are unfamiliar with the object or if halfway through your reach I flip the object on its side or push it closer to you. These are the conditions that require analysis of visual details about the particular shape and location of the object, as opposed to its "what" properties, and these are the conditions that are most impaired in optic ataxia.

Observations like these have led to the view, championed most prominently by psychologists Melvyn Goodale and David Milner, that there are in fact two visual systems, a ventral system for perception (recognition) and a dorsal system for action.[21] If you think about it, it almost couldn't be otherwise. The functional necessities for recognition versus action are very different. For recognition, visual information needs to make contact with neural codes for stored memories of the meaning or semantics of an object (is it a cup or a cat?) and requires abstracting over particulars. For action, visual information needs to make contact with motor programs for controlling the body and this requires detailed analysis of the particulars of the visual scene in relation to the body. To the extent that the neural codes are different for abstracted semantic "what" versus motor control "how," the neural circuits for recognition versus grasping must be different.

By analogy, if you focus your smartphone camera lens on a QR code (those square "bar codes") using the camera app, the input from the photo sensor appears on the screen as a pattern of color-coded intensity values and retains information from shading, glare, distortion, or occlusions from the angle of the photo. If you point your phone at the same pattern using a QR code reader app, the image is processed in a very different way. Once the photo sensor has done its work, certain portions of the code are used for finding the edges of the pattern, others for alignment, others for format (text, a link to a website, Chinese characters), and others for the content itself; the details of shading and highlights, orientation, distortions, or occlusions are discarded or corrected. Then, rather than displaying or storing the image, the processed data are used to do something, like display text or launch the web browser and link to a site. The very same input is used to do very different things with very different outcomes. To do this you need different computational processes. Importantly, the camera app isn't very good at QR code reading and the QR reader app isn't good at photo reproductions. Why? Because the computations involved are tailored to the goals, the function of the app.

The same is true in the brain. Even though the retina registers the same information from a given cup no matter what, the brain circuits (computations) that get involved up in cortex depend entirely on the neuro app that is invoked by whatever task is at hand. And it turns out that the recognition neuro app involves a different circuit from the motor-control neuro app. These are the ventral *what* and dorsal *how* streams.

These two streams are not entirely independent. In fact, they necessarily interact on some level: we might want to reach for a bit of rope, but we would most often want to avoid reaching for a snake. We sometimes need to recognize the object before deciding what to do with it. But this doesn't change the fact that the computations involved in recognizing and reaching have to be different. The features of the object that are useful for recognizing and reaching are not the same, and the endpoints of the processes are very different (a long-term memory representation of the object meaning or category versus a code for controlling limb movements). The fact that we too often react to sensory input without full understanding—*Act, don't react!*—tells us that the dorsal action circuit does not depend completely on the output of the recognition circuit. It can act on its own.

From an evolutionary standpoint we can make sense of why we should have two systems, one that (re)acts and one that recognizes. If you're not a very complex organism it might be enough to sense something, like an uncomfortable poke, and simply move until you eliminate it. This could improve your chances of survival and works well while lounging on the couch watching TV. But the world is more complex. In many circumstances it helps to react differently depending on what is behind the sensation—it's important not to mistake the remote for a Snickers bar—or to not react at all. Thus, a second sensory system tuned to recognize objects by linking what is being sensed with stored memories of similar objects or past events confers even more survival value. It doesn't replace the first, reactive, system; we still need to be able to react. Instead it drastically increases the *flexibility* of our response. We can act now, later, or never, depending on how we understand the object or situation.

Dorsal *what* and ventral *how* streams exist in the auditory domain as well. Just like there are at least two different things we need to be able to do with the visual image of a cup, there are at least two different things we need to do with the acoustic pattern of the word *cup*. On one hand we need to understand what the acoustic squiggle of chirps, honks, and swooshes *mean* (and their variants across individual voices and accents), the ventral stream's job. On the other hand we need to be able to reproduce a similar acoustic squiggle with our own vocal tracts, the dorsal stream's job. As with visual agnosia and optic ataxia in the visual streams, the functions of the two auditory streams also dissociate following brain damage.[22]

So now we are finally in position to see why the proposed action understanding function of mirror neurons doesn't fit the dorsal-ventral principle. Mirror neurons are, by all accounts, part of the motor system. As such, they are part of the dorsal, *how*, stream. The anomaly, then, is this: what's a *what* function doing in the *how* stream? The features that are relevant for motor control are not the features that are useful for recognition, as we've seen. This is even more puzzling in the context of the neural neighbors of mirror neurons, so-called canonical neurons. Like mirror neurons, these cells respond both to sensory stimulation (the visual perception of objects) and during action execution. The interpretation of these cells by the Parma group followed standard dorsal stream function, specifically that canonical neurons code a "vocabulary of motor acts and that this vocabulary can be accessed by . . . visual stimuli;"[23] that the canonical neuron system is critical for "learning associations, including arbitrary associations between stimuli and [motor] schemas;"[24] and that this is a "'pragmatic' mode of processing, the function of which is to extract parameters that are relevant to action, and to generate the corresponding motor commands," as opposed to "'semantic' analysis [which is] performed in the temporal lobe."[25] Thus, canonical neurons respond to objects not because the neurons are part of a mechanism for understanding objects, but because objects have features relevant to guiding action. Yet, nearby mirror neurons, which show the same response properties as canonical neurons (they respond both during execution and observation), are endowed with a ventral-stream-like semantic function. This is an anomaly, both functionally and anatomically.

It could be that mirror neurons really are a functional outlier in the grand scheme of cortical organization. But if this is true, we're going to need exceptionally compelling evidence. It is yet to arrive.

ANOMALY 8: SIMULATION AND "GOAL ASCRIPTION" AS A ROUTE TO UNDERSTANDING

THE INITIAL excitement over mirror neurons centered on cells that showed a congruence between the motor actions they seemed to code and the observed actions they responded to, so-called *congruent mirror neurons*. The Parma group comments:

> Most importantly, mirror neurons show a very good congruence between the effective observed and the effective executed action. This visuomotor congruence has prompted the idea that the basic function of mirror neurons consists in understanding actions made by other individuals.[26]

By way of reminder, this congruence enables "understanding" in the following way, according to Gallese et al.:

> When an individual emits an action, he "knows" (predicts) its consequences. This knowledge is most likely the result of an association between the representation of the motor act, coded in the motor centres, and the consequences of the action. Mirror neurons could be the means by which this type of knowledge can be extended to actions performed by others. When the observation of an action performed by another individual evokes a neural activity that corresponds to that which, when internally generated, represents a certain action, the meaning of it should be recognized, because of the similarity between the two representations.[27]

And per Rizzolatti and Craighero:

> Each time an individual sees an action done by another individual, neurons that represent that action are activated in the observer's premotor cortex. This automatically induced, motor representation of the observed action corresponds to that which is spontaneously generated during active action and whose outcome is known to the acting individual. Thus, the mirror system transforms visual information into knowledge.[28]

The complication here is that the movement itself—the thing that is supposedly mirrored or simulated and automatically leads to a prediction of the outcome—is ambiguous. If we are sitting in a café and I reach toward the table between us with my hand oriented as if to grasp an upright cylinder, you can simulate my action and maybe predict, from the orientation and anticipatory sizing of my hand, that I am going to grasp something of a particular size and orientation. That's potentially useful information. By simulating my movement, you could come to understand that I am about to grasp a cylinder. Three questions come to mind, though:

1. *Do you need to simulate my movement to "understand" this much about my action?* No, you could do it by pure association between hand shaping and object size. My dog loves to play fetch. Through much experience with me throwing the ball for him, he can now easily and reliably predict the direction the ball will fly based on the orientation and trajectory of my wind up. Since he can't throw, he can't be simulating the movement to "understand" what might happen. He does it by association. You can too.

2. *Are mirror neurons doing this kind of prediction, that is, using hand-shaping information to predict what kind of object might be the target of the grasping?* Not in any simple sense. A variant of motor simulation, inspired by the description of the mechanism quoted above, would involve activating the corresponding motor program to the one you are observing and then seeing what consequences become activated. Pinch-like "precision grip" extended forward→Ah, small-object grasping! Whole-

hand "power grip" extended forward →Ah, large-object grasping! The process starts by simulating the actual movement and then proceeds, but this isn't how mirror neurons behave. Mirror neurons don't simulate the movements themselves in any simple way. They simulate the movement only in the appropriate context. A precision grip action is not simulated by a mirror neuron unless the object of the action, the small object in question, is present. So the process can't be: simulate the action→ see what consequence becomes activated. It has to be something like: Does the observed action have a potential target? If it does, then simulate.[29] But once the system has evaluated whether the action has a potential target, it's not at all clear that the simulation adds anything in terms of "understanding" the immediate goal of the action. If simulation allows us to understand something like, *that action will result in grasping of the raisin*, but you need to identify that the action is raisin related before you simulate it, then you already know the raisin is about to be grasped prior to simulating. But perhaps mirror neurons are providing a deeper level of understanding via their simulation of observed actions, which leads us to question 3.

3. *Do mirror neurons provide information about the* intention *behind the action?* This could be extremely useful information. After all, if we really want to understand an action, we need to go beyond the recognition that, say, a small object is being grasped. We need to know *why* it is being grasped. This is exactly the theoretical direction that Rizzolatti and his colleagues have moved toward in recent years. A key finding in this respect was reported in a 2005 paper published in the journal *Science* under the title "Parietal Lobe: From Action Organization to Intention Understanding."[30]

The experiment adopted an extremely clever design. Two monkeys were trained to grasp food or solid nonfood objects and place them either in their mouths (the food) or in a small container (the solid objects). For most of the experiment, the container was situated on the table to the left or right of where the food/object was placed by the experimenter. For another portion of the experiment involving fewer trials, the container was situated right next to the monkey's mouth, as if it were sitting on his shoulder. This allowed the team to control for possible differences in the movements between the two action goal conditions, grasping to eat versus grasping to place. Neurons were recorded from the parietal lobe region previously found to contain mirror neurons.

In the first phase of the experiment, the investigators studied 165 neurons, all of which were active during the execution of grasping actions. Thirty-six percent of the neurons responded equally well during grasping to eat and grasping to place actions. But a majority, 64 percent, varied in their response depending on the goal. Roughly three-fourths of the goal-selective cells preferred grasping to eat (no surprise there!) while a quarter preferred grasping to place (somebody's got to do it). A subset of cells (18) were studied in all three conditions, eating, placing near object, and placing near mouth. Fourteen of these cells preferred grasping to eat and discharged less when the object was placed in the container near the food, which is not terribly surprising because the movements were very different. But the same 14 cells also fired less vigorously during grasping-to-place actions (compared to grasping-to-eat actions) when the container was right next to the monkey's mouth. This is important because the physical movements in the two actions (eating and placing) were nearly identical; the cells did not seem to be coding the movements themselves, but rather the *goals* of the actions. Similarly, the 4 neurons that preferred grasping to place responded equally well to placing near the object as placing near the mouth. "The main factor that determined the discharge intensity was, therefore," the authors wrote, "the goal of the action and not the kinematics" (p. 663); kinematics refers to movement mechanics, the pattern of muscle contraction and relaxation that achieves the movement.

These are interesting results but the cells under study weren't necessarily mirror neurons because only motor execution responses were measured. In the next phase of the study the group set out to see whether mirror neurons behave in a similar, goal-dependent fashion. Forty-one mirror neurons were identified, cells that responded not only during grasping but also while the monkey observed the experimenter performing the same grasping-to-eat and grasping-to-place tasks. Would the mirror neurons also show a goal preference? The answer was yes for a majority of them (roughly three-fourths, or 31 of the 41 cells). Of these, most preferred, you guessed it, eating (23 cells). A subset of the goal-selective mirror neurons (19) was evaluated for congruence between their execution and observation selectivity (do they prefer the same type of movement during execution and observation?). Most (16) showed congruent preferences for action and observation. Thus, mirror neurons are *goal*-selective.

This goal-, not movement-, oriented effect isn't a fluke. Another oft cited paper reports a similar result in frontal motor area F5, namely that the cells respond to action goals, not the kinematics of the movement, although the setup was rather different.[31] In addition, human imaging studies have demonstrated similar effects showing that the mirror system is not responding to the movement per se but rather to the context in which the movement is executed.[32]

But let's return to the original study. The neuronal property documented in these parietal neurons, the goal-selective response, "allows the monkey to predict the goal of the observed action and, thus, to 'read' the intention of the acting individual." Or so the story goes. A closer look at the results raises the same question we had about predicting the target of a simple action. Does simulating the movement augment the monkey's ability to predict the goal?

Here is a telling nugget of information about the setup of the experiment in the action observation conditions. The presence or absence of the container fully predicted the human experimenter's action. When the container was present, the subsequent action was always placing; when it was absent, the action was always eating. And the preferential response of the parietal cells, to placing versus eating, was evident during the experimenter's reach *toward* the object, which occurred in both conditions and therefore provided no cue itself regarding the end goal of the movement. How did the monkey's neurons predict the goal? It was the *nonmotoric* context, not the observed movement. The movement itself was completely ambiguous between placing and eating during the same time frame that the mirror neurons were "predicting" the goal. The information simply was not in the action.

The researchers were fully aware that the goal was predictable from the context. In fact, they designed the experiment that way. An alternative to this design might have been to leave the container in place during both placing and eating trials. To visualize this, put yourself in the monkey chair and imagine an experimenter with a jar on his shoulder and a piece of food on the table. He reaches for the morsel and grasps it. To this point you can't tell what he's going to do. Then he begins an arm movement back toward his head. Still it may be impossible to predict the outcome. Finally, the hand passes the mouth and drops the food in the jar. In this scenario, the movements are identical until the very last second, making it impossible to predict the outcome before it actually happens. So, if this setup were used in the monkey experiment, there would have been no way for the cells to respond differentially until the placement of the food was accomplished and by that time there would be nothing to predict. By removing the container on eating trials and including it on placing trials, the experimenters removed the inherent ambiguity in the movement itself and made it possible for the cells to show what they knew and when they knew it. Given that the outcome predictability is coming from the setup and not the movement, we might ask, how is it possible to conclude that the mirror neurons are predicting the outcome by motor simulation when there is nothing in the movement that allows for such a prediction?

Here's how the authors of the study explained it. Motor acts are often chained together. We reach for a coffee cup and then bring it to our lips; we toss a ball in the air and then hit it with a racquet. The fluidity with which we execute such sequences of movements suggests that they are linked in the neural code: activation of one segment of the movement chain triggers the activation of the next. This is a fairly well established idea, one well illustrated by the difference between trying to type an unknown word (one letter at a time) versus your own name (the sequence just flows off your fingers). Applying this idea to their experimental finding, the authors proposed that the parietal mirror neurons that discriminate grasping-to-place from grasping-to-eat actions are coding these larger motor chains, not the individual movements. Therefore, when a movement occurs in a context that is most consistent with one chain versus another, the system can activate the appropriate chain, simulate the movement *chain,* and from this predict the outcome.

This is an elegant idea, but it doesn't solve the problem, as philosopher Patricia Churchland has noted.[33] To know which chain to activate, you have to know ahead of time whether the movement is likely to end in placing (a container is present) or end in eating (no container is present). And if you have to know ahead of time, what does simulation add to your knowledge? It's the same problem encountered previously regarding regular mirror neurons and the need to have an object present. Hungarian cognitive scientist Gergely Csibra summed up the problem eloquently:

> [There is] a tension between two conflicting claims about action mirroring implied by the direct-matching hypothesis: the claim that action mirroring reflects low-level resonance mechanisms, and the claim that it reflects high-level action understanding. The tension arises from the fact that the more it seems that mirroring is nothing else but faithful duplication of observed actions, the less evidence it provides for action understanding; and the more mirroring represents high-level interpretation of the observed actions, the less evidence it provides that this interpretation is generated by low-level motor duplication.[34]

Another frequently cited paper published in 2001 by many members of the original Parma team along with some collaborators illustrates this tension directly.[35] The standard mirror neuron experiment was carried out, including action execution as well as action observation tasks, but with a twist. The action observation task had some added conditions. In one, the monkey observed the actions in full view as was done in previous experiments. In another condition, however, after the monkey was shown the object sitting on a platform, an opaque screen was placed in front of the object hiding it from view. The experimenter then reached behind the screen for the object as before. What would mirror neurons do in the hidden condition? Mirror neurons don't respond during pantomimed actions; the object of the action has to be present. So would mirror neurons shut down without the visual context? Or would mirror neurons respond because the monkey "knows" that the object is there? You can guess the outcome based on the title of the report: "I Know What You Are Doing: A Neurophysiological Study." Indeed, a number of mirror neurons persisted in firing even when the object was hidden. (A control, no-object condition was also tested in which the monkey was shown an empty platform, the screen was lowered, and then a reach was executed as if the object were there; mirror neurons did not fire in that situation.)

As with the experiments described above, this result was construed as evidence that mirror neurons were tracking the goal or meaning of the action, not the physical features of the action itself.[36]

The "tension" that Csibra notes is made clear by the Parma group and their collaborators in the discussion section of the report:

> In order to activate the neurons in the hidden condition, two requirements need to be met: the monkey must "know" that there is an object behind the occluder and must see the hand of the experimenter disappearing behind the occluder. It appears therefore that the mirror neurons responsive in the hidden condition are able to generate a motor representation of an observed action, not only when the monkey sees that action, but also when it knows its outcome without seeing its most crucial part (i.e., hand-object interaction). (p. 96)

If the monkey already knows the outcome, what is the point of simulating it with a motor response?

It would seem that the mirror neuron theory of action understanding is caught between a rock and a hard place. Simulating doesn't help unless you know the outcome (because the movements alone are ambiguous) and if you know the outcome there's no point in simulating. But Rizzolatti and Italian philosopher Corrado Sinigaglia have attempted to rescue the theory (unsuccessfully in my view) in a direct response to this type of critique.[37] They argue that the mirror system contains two kinds of sensorimotor transformations, one that indeed mirrors the details of a movement directly and one that mirrors the goal. They then state further that direct matching of movements— literal simulation, which motivated the whole theory in the first place—is "devoid of any specific cognitive importance *per se*." "By contrast," they continue immediately, "through matching the goal of the observed motor act with a motor act that has the same goal, the observer is able to understand what the agent is doing." So there are two processes, one that mirrors movements and doesn't play any important role in understanding (rock) and another that mirrors the goals of the action and therefore isn't driven by motor *simulation* per the arguments above, which contradicts the fundamental thesis of the whole motor simulation enterprise (hard place).

One last bit of data bears on this discussion because it questions whether the mirror system mirrors goals at all. The study involves fMRI data collected from a group of macaque monkeys.

Getting a monkey to submit to an MRI scan is an impressive accomplishment, but it is possible and it provides a nice complement to single unit research because it affords a wider angle perspective on the neurophysiological response. To accomplish this feat, Rizzolatti's Parma lab teamed up with a group of scientists at Belgium's Katholieke Universiteit Leuven.[38] They trained five monkeys to undergo fMRI while observing videos that showed various actions including an experimenter grasping different objects (as is typical in mirror neuron recording studies), a close-up view of just an arm grasping objects, a robot hand grasping objects, a human hand pantomiming an object grasp, and objects moving around on their own. Static video frames and scrambled videos that look like dynamic grayscale patterns were used as controls.

As expected, observation of actions activated F5 as well as other frontal regions relative to the control conditions. One interesting result was that the subportion of F5 known to contain mirror neurons, F5c, responded to the full-frame action videos but not the zoomed in, arm-only actions. The entire context, white-coated experimenter included, was required to drive action-related activity in the "mirror neuron region." Two other regions of F5 (F5p and F5a) along with another area altogether, area 45B, responded both to the full-view and the zoomed-in action videos.

This finding raises a theoretical problem with respect to the goal-directedness of the response of mirror neurons. In both the full-view and arm-only conditions, the goal of the action was the same: to grasp the object. Yet mirror neurons apparently did not respond to both of these conditions, which suggests that they are not responding to the action goals but something more idiosyncratic to the particular action situation.

Results from the other observation conditions (the robotic and pantomimed actions) are also revealing. The mirror neuron region (F5c) did not respond in any of the other conditions. However, a different F5 region (F5a), responded not only to goal-directed human actions but also to goal-directed actions by a robot hand and to non-goal-directed (pantomimed) actions. Although it has not yet been studied directly using single cell recording methods, the authors of the study point out, quite correctly, that given the location of F5a within premotor cortex it is likely that the cells that respond to the observation of robotic and pantomimed actions also have motor properties. From this we can *tentatively* infer the existence of a new class of mirror neurons that respond both during action execution by the monkey and the observation of robot and pantomimed actions. The authors suggest that this response reflects the representation of the abstract action. And this, it is further speculated, links the monkey mirror system to language, a system that is virtually defined by abstraction:

> Thus, it is plausible that the transition, in the monkey frontal lobe, from context-dependent descriptions in F5c to more abstract descriptions in F5a . . . represents the ancient prelinguistic basis from which the abstract description of an action, necessary for language, evolved. (p. 336)

This is a fairly dramatic claim for a study that seems to show (i) that the mirror neurons that have been studied since 1992 are not goal directed after all but are context dependent in a much more idiosyncratic way and (ii) that there is another, yet-to-be-studied region that provides the basis for the evolution of language by virtue of the fact that it responds to actions independently of the object-context or the agent (i.e., it is not coding goals) and therefore fails to specify "*information critical for understanding action semantics, i.e., what the action is about, what its goal is, and how it is related to other actions*"[39]—to quote from the introduction of the very same paper.

In retrospect, there was evidence in the earliest reports that mirror neuron responses were not predominantly goal directed, but in fact rather more idiosyncratic. For example, the 1996 monkey mirror neuron publication reported that 30 of 47 mirror neurons tested were movement-direction selective, that is, they responded more vigorously depending on whether the observed action moved from left to right or right to left.[40] The goal certainly doesn't change depending on movement direction, so why the selectivity?

Looking at the collection of results summarized here we have to conclude that if being goal directed is critical for understanding action semantics, then it is not clear at all that mirror neurons have the right properties to enable action understanding.

WHAT DO ANOMALIES TELL US?

THE EXISTENCE of anomalies for a theory does not necessarily prove that the theory is wrong. Good theories almost always face empirical challenges. In some cases, the data turn out to be invalid or misinterpreted. In other cases, relatively minor tweaks to the theory can account for the new facts. But in some cases, the anomalies may ultimately point to a different theoretical point of view altogether; Earth revolves around the Sun. In all cases, anomalies demand a closer look at the facts and the theoretical assumptions.

None of the anomalies that I pointed out here, individually, should lead to a complete abandonment of the mirror neuron theory of action understanding. Indeed, as we've seen, various tweaks to the model have been proposed to accommodate some of these facts. However, the sheer number of anomalies is a significant concern and one that leads us to consider other sources of data to test the theory, as well as to explore alternative explanations for mirror neuron behavior. This is the topic of the next chapters. We start with the very behavior that inspired the mirror neuron theory of action understanding and the one that it seems most bent on ultimately explaining: language.

5

Talking Brains

T O UNDERSTAND why language is often a centerpiece in mirror neuron theorizing—indeed a critical test bed for the theory—we have to trace the story back almost four decades before the Parma team's first reports on the monkey cells. A good starting place is June 16, 1956, the date that American psychologist Alvin Liberman presented a paper titled "Some Results on Research of Speech Perception" at the Conference on Speech Communication at the Massachusetts Institute of Technology. Liberman's investigations sought to uncover the acoustic features associated with the perception of phonemes, the individual speech sounds of a language that correspond roughly (very roughly in English) to alphabetic letters; consonants were a major focus of this effort. After describing the general approach of his laboratory he started the meat of his discussion with a caveat:

> As a matter of convenience, I should like at the outset to divide the consonant cues into three classes, and to make this division according to where and how the sounds are produced. I am, of course, embarrassed to introduce a discussion of acoustic cues by classifying them on an articulatory basis. However, we find

here, as we so often do, that it simplifies our data quite considerably to organize them by articulatory criteria. We certainly do not mean to imply by this that there are no acoustic differences among our classes, but only that it is hard to characterize these differences very simply in acoustic terms.[1]

The roots of the mirror neuron theory of action understanding can be found in Liberman's caveat. A little scientific context helps illuminate how.

Liberman's group at Haskins Laboratories—a world renowned center for speech and language research—became puzzled by the observation that the rate at which phonemes occur in spoken language seems to exceed the temporal resolving power of the auditory system. For example, in one of their most influential papers, "The Perception of the Speech Code," the Haskins team estimated that conversational speech can easily contain 15 phonemes per second and listeners can tolerate more than 20 or 25, as the hushed audio "small print" at the end of some radio advertisements attests. By comparison, if you crammed 15 notes of Happy Birthday into one second, the tune would be completely unrecognizable.

How does the auditory system successfully process a dozen or so discrete speech units every second? It doesn't. Liberman and his colleagues discovered that unlike printed letters on a page or the notes of a simple tune, phonemes are not discrete elements in the acoustic speech stream. Rather, information about individual phonemes is smeared out in time and overlaps with neighboring sounds. If letters were written like phonemes are spoken, they would look more like this:

Phonemes are not discrete elements

And even this is cheating a little bit because I've left gaps between the words. Speech doesn't do this. The phonemic smearing crosses word boundaries as well.

One of the earliest hints that this was the case came from efforts to cut up tape recordings of speech into phoneme-sized chunks corresponding to the different vowels and consonants, a kind of audio alphabet, and then recombine them to play back new words. Such attempts failed miserably. The resulting speech was largely unintelligible.[2] Liberman and colleagues adopted a more systematic approach to identifying the acoustic building blocks of speech by using simple consonant-vowel (CV) syllables generated on a pattern playback machine, an early speech synthesis device, which allowed for tight control over the sounds. They began lopping off increments of an acoustic syllable such as *da* from "right to left" to see if there was any point at which they could isolate just the consonant sound. There was no such point. "At every instant," they wrote, "[the acoustic signal is] providing information about the two phonemes, the consonant and the vowel—that is, the phonemes are being transmitted in parallel."[3]

The acoustic speech signal transmits information about phonemes in parallel because we articulate speech sounds with our vocal tract in parallel. Say the words *key* and *koala* and notice the difference in your lip position when you make the /k/ sound: in one case (*key*) your lips are drawn back and in the other (*koala*) they are rounded. You are anticipating the articulation of the following vowel even before you articulate the consonant. This phenomenon, called *coarticulation*, smears acoustic cues for neighboring phonemes.

The discovery of parallel transmission of phonemic information in the speech signal left in its wake a profound problem in speech perception that still hasn't been resolved. It's known as the *lack of invariance problem* and it led to the formulation of the motor theory of speech perception and thus indirectly the mirror neuron theory of action understanding. Here's the core of it. No consistent ("invariant") acoustic cues correlate with the perception of (some) phonemic units. This is most evident in the case of so-called stop consonants, sounds like /b/, /d/, /g/, /p/, /t/, and /k/. These sounds are articulated by a complete stoppage of airflow within the vocal tract followed by an abrupt release of air.

Consider the /b/ sound in *about*. To make the /b/ sound, your mouth transitions from the open /a/ position to a closure of the lips, which stops the airflow. A slight amount of pressure is allowed to build behind the lips and then they are opened, releasing the air in a burst. By closing the vocal tract in different positions, we can generate different sounds: lips = /b/ and /p/, tip of the tongue on the roof of mouth behind teeth = /d/ and /t/, and back of the tongue on the roof of mouth = /g/ and /k/. The difference between the two sounds that result from closure at the same place in the mouth (/b/ versus /p/, for example) is the *voice onset time* of the following vowel: whether we voice (vibrate our vocal cords) at the same moment that we release the burst of air or whether we wait 20–30 milliseconds before voicing the vowel.

Say *bat* and *pat* as slowly as you can out loud. Notice that to make the /b/ sound in *bat* you have to start voicing the /a/ sound right when you release your lips, whereas you can make the /p/ sound simply by releasing an "unvoiced" puff of air. While it is fairly straightforward to describe how stop consonants are articulated, Liberman and colleagues noticed something odd when they looked in detail at the acoustics of these sounds. The problem is illustrated in this figure:

Lack of invariance problem

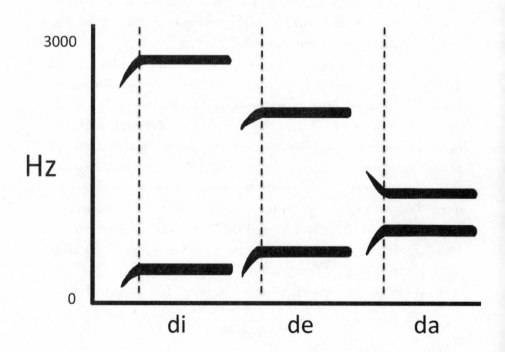

The figure shows an idealized graph of the acoustics of three syllables, *di*, *de*, and *da*. This type of graph is called a spectrogram because
it shows the spectrum (the frequency range of wavelengths) in the
acoustic signal over time. Recall that sound is transmitted by waves
of air pressure, like ripples in a pond. These waves can vary in length,
the distance from one peak to the next, which is also called the
wave *frequency* and is measured in Hertz (Hz). Our auditory system
translates differences in wave frequency into differences in perceived
pitch: high frequencies sound high pitched and low frequencies sound
low pitched. For reference, the (youthful) human auditory system
can hear frequencies in the range of about 20–20,000 Hz, although
a smaller range from about 100–10,000 Hz is most audible. Speech
frequencies range from roughly 100–5000 Hz and middle C on a
piano is 261.6 Hz.

Most natural sounds, including speech, are a mix of many different frequencies, like complex musical chords. To visualize the range
of frequencies (the spectrum) in a sound, we can graph frequency on
the vertical axis and time on the horizontal axis. Such a spectrogram
shows us which frequencies are present in the sound at which points
in time.

Returning to the figure, notice that each syllable has a different
acoustic pattern. This is expected because each syllable sounds different. But look more closely. Focus first on the flat portions to the
right of the dashed lines. These parts of the graph correspond to the
different vowels and again we see what we expect: different vowel
sounds are associated with different acoustic patterns. Now focus on
the swoopy portion to the left of the dashed lines, which is the part of
the pattern that is needed to distinguish one consonant from another.
Here again each pattern is different, especially in the top band, yet
what we hear in each case is *not* different: we hear the same /d/ sound.
This is the lack of invariance problem that Liberman noticed: different acoustic patterns (as opposed to an invariant pattern) result in our
hearing the exact same speech sound. It didn't make sense.

As is evident in the quote at the beginning of this chapter, Liberman and colleagues noted, however, that there *is* invariance, a consis-

tent pattern, in the way a given phoneme is articulated. Say, *di-de-da*. In each case there is a commonality in the way you make the /d/ sound: in each case you stop airflow with the tip of your tongue on the roof of your mouth. This consistency, coupled with the lack of consistency in the acoustic pattern(s) associated with that phoneme, spawned Liberman's radical proposal: we don't perceive speech *sounds*, we perceive speech *gestures*. We don't recognize a /d/ by its acoustic pattern, we recognize it by the way it is *articulated*. This is the motor theory of speech perception.

If you are little confused about how it is possible to recognize speech *gestures* from an *acoustic* speech signal, you are not alone. That part of the theory was never actually worked out. It was merely assumed that because listeners are also talkers, they have the specialized, inside knowledge to somehow recognize gestures from acoustic patterns. This part of the theory always bothered me. If a particular acoustic pattern is effectively ambiguous, I wondered (and continue to wonder), then how do you know which gesture it goes with? But the Haskins group was thinking bigger. If the theory were correct, the lack of invariance problem would be solved. What was needed, then, before trying to work out all the minute details, was evidence that the approach was on the right track. This evidence soon arrived in several forms.

One was the discovery of *categorical perception*. Having observed that the acoustics of a speech sound could vary while the percept of the phoneme remained steady, Liberman and his Haskins team asked whether listeners could hear acoustic variation at all: are listeners able to detect the fact that the /d/ in *di*, *de*, and *da* is not acoustically the same? Using their pattern playback machine they generated what they knew to be good examples of three different stop consonants in the context of a given vowel, *be*, *de*, and *ge*, which correspond roughly to spectrograms 1, 6, and 11 in the figure below. They then generated a *continuum* of "in between" stimuli in which the swoopy transition part of the top frequency band was changed in small steps from one phoneme exemplar to the next, as pictured (they actually used 14 steps from /b/ to /g/). They were curious how listeners would perceive the in-between sounds.

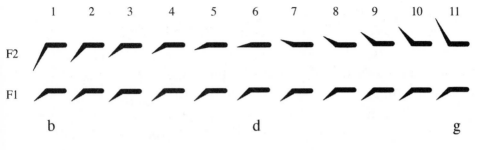

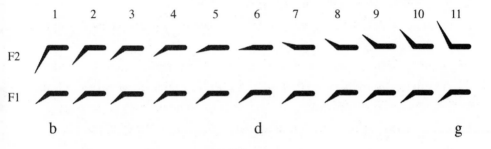

Study participants were asked to perform two tasks. In one, they simply listened to the stimuli presented several times in random order and labeled the sounds as either /b/, /d/, or /g/. The investigators found a fairly abrupt transition between the stimuli labeled as one category—*be*, *de*, or *ge*, compared to the next. Although the acoustics varied continuously in even steps, subjects tended to lump each stimulus into one phonemic category or another.

This perhaps is not a terribly surprising result. After all, subjects were *asked* to do this—to lump the stimuli into one speech category or another. Consider the analogous task of lumping a continuous range of the height of individuals into tall or short categories. Some cases are clearly members of one category or the other, and some individuals are in between. But if you are asked to categorize, you decide on a rough category boundary and call everyone below that height short, above it tall, with maybe a little slop around the boundary, effectively ignoring intracategory variation.

This is where the second, critical task comes in. Subjects first heard two different stimuli that could be one, two, or three acoustic steps apart (analogy: you see someone who is 5'5" followed by someone who is 5'6", one step; or 5'7", two steps; or 5'8", three steps apart). Then they heard a third stimulus that matched either the first or second stimulus and they had to indicate which one it matched. Their decision could be based on any acoustic cue in the stimulus; the phonemic category (/b/, /d/, /g/) wasn't necessarily important. The idea is that if subjects can *hear* the acoustic variation, they should be able to discriminate the stimuli even if they belong to the same phonemic category. By analogy, we can ask whether you can detect height differences between people who you categorized as short, say, someone who is 4'11" versus another who is 5'2", or whether you perceive "short people" as being one height.

In their 1957 paper the team reported that subjects could *not* discriminate among sounds that were within a phonemic category as well as they could discriminate among sounds that were equal steps apart but crossed their own phonemic category boundary.[4] The authors concluded that the phonemic categories of a language influence the way we perceive acoustic speech information to the extent that our perceptual systems are for the most part blind to acoustic variation between speech categories, such as /b/, /d/, and /g/. This is categorical perception, and the 1957 report suggested that it is characteristic of how humans perceive speech. Later research seemed to show that categorical perception was specific to speech sounds; it didn't apply to similar acoustic patterns that were not perceived as speech.[5]

This is a dramatic finding. If it held in our height perception analogy it would mean that we perceive only two heights, short and tall, and we can't tell the difference on a basketball court between a 6'3" point guard and a 7'0" center.

Categorical perception and its presumed specificity to speech fit the motor theory of speech perception perfectly. The auditory system generally is set up to analyze acoustic features in a continuous manner, so the story goes, and this is what we use when perceiving the pitch and timbre of a cricket chirp or a car horn. But Liberman's research showed that our perception of speech sounds does not conform well to the world of continuous auditory features. This observation could be understood if our speech percepts were derived via a different mechanism, one that codes speech categorically. And where does such a code live? In the motor system, with the (invariant, categorical) neural codes for controlling speech gestures.

Another important source of evidence claimed to support the motor theory came from an accidental discovery in the late seventies by psychologist Harry McGurk and his research assistant John Mac-Donald at the University of Surrey in England. They were studying infants' ability to perceive speech sounds and were using both auditory and visual speech signals (videos of someone speaking). They ended up mixing the audio of the syllable *ba* with a video of a talker saying *ga*. When they watched the dubbed clip, they heard something different still, a *da*. Once they confirmed that it wasn't a dubbing error but rather their brains confabulating a syllable that wasn't in the audio or video signals, they ran a formal experiment to establish the effect in a group of subjects and published their results in 1976 in a report titled "Hearing Lips and Seeing Voices" in the journal *Nature*.[6] Entire scientific careers have been spawned by the accidental discovery of the *McGurk effect* as it is most commonly known (or *McGurk-MacDonald effect,* more accurately) and it remains a topic of much study today.

Why does the McGurk-MacDonald effect offer evidence for the motor theory of speech perception? Because it seems to show that information about the motor gesture, perceived in the visual signal, influences speech perception. And not only that, the particular percept that results from mixing a visual *ga* and audio *ba* can be understood in motor speech terms: /b/ sounds are made in the front of the mouth by closing the lips together whereas /g/ sounds are made in the very back of the mouth by raising the back of the tongue so that it touches the velum or soft palate. The /d/ sound falls in between /b/ and /g/ in terms of how it is articulated in the mouth: the tip of the tongue is raised to the alveolar ridge, just behind the teeth. It's as if the percept of *da* is an average of *ba* and *ga, in speech gesture space.*

At first blush a decent collection of circumstantial evidence supports the motor theory of speech perception. But as with mirror neuron theory, as scientists investigated more deeply, anomalies started to emerge. With respect to categorical perception researchers in the 1970s discovered that creatures devoid of the ability to produce speech—namely, chinchillas and one-month-old infants—nonetheless exhibited categorical perception of speech sounds.[7] The infant study, carried out by a team led by developmental psychologist Peter Eimas at Brown University, was the first to make a splash with their paper published in the journal *Science* in 1971.

Because infants don't talk and have trouble following instructions, the team needed a method for assessing the perceptual talents of their subjects. They discovered that sucking is the key to a baby's mind: when infants are interested in something, they suck faster. So Eimas and colleagues rigged a pacifier that could record sucking rate and then presented speech sounds to one- and four-month-old infants. After a baseline sucking rate was established, a syllable (e.g., *ba*) was played over and over for five minutes. During the first 2–3 minutes of listening, the infants roughly doubled their sucking rate. After that, sucking tempo slowed so that by the end of the five-minute period, sucking rates approached the baseline level. The infants had habituated to the repetitive syllable. At this point, and without interruption, a different syllable (e.g., *pa*) was repeatedly presented for the next four minutes. Infants' sucking rate went back up, indicating that they perceived the difference between the two syllables. Another condition provided the critical test. After presenting one syllable for five minutes and observing habituation in the infants' sucking, the researchers presented a second stimulus that differed acoustically from the first—as different as the *ba* and *pa* stimuli were acoustically—but was within the same phoneme category as judged by adults. Neither the one- or four-month-old infants noticed the within-category acoustic change. Babies perceive speech sounds categorically.

In 1975, four years after the infant study was published, Patricia Kuhl and James Miller reported, also in the journal *Science*, that chinchillas exhibit human-like categorical tendencies when perceiving speech. Chinchillas were selected for this study because their auditory systems are similar to humans' in terms of the range of frequency sensitivity. A group of animals were trained to respond differently to syllables with /t/ versus /d/ as the first phoneme. Training was carried out using syllables recorded by four different talkers and in three different vowel contexts (*ti, ta, tu; di, da, du*). Impressively, the critters learned to recognize the difference between /t/ and /d/ sounds, and the learning generalized to new instances of the same syllables, including synthesized versions. The researchers then tested the animals' ability to categorize speech by presenting a continuum of sounds between *ta* and *da*. The chinchillas tended to lump the stimuli into one category or the other with a boundary that was the same as a group of human English-speaking listeners. Clearly, motor speech competence is not the basis of categorical perception.

A whole range of studies also challenged the uniqueness of categorical perception to speech. As perceptual scientists explored other types of stimuli, they found evidence for categorical perception of facial expressions, colors, musical chords, the sound of slamming doors, birdsong (in birds), and sound frequency (in crickets).[8] The tendency to categorize our sensory environment appears more the rule than a unique feature of speech perception.

Research in the 1980s by American psychologist Dominic Massaro further questioned categorical perception as an empirical pillar of support for the motor theory of speech perception by showing that while we do tend to categorize, human listeners are actually quite capable of perceiving within category acoustic variation in speech sounds. Massaro and his collaborator Michael Cohen performed a similar experiment to Liberman's *b-d-g* continuum study but instead of asking their experimental participants to discriminate or identify the sounds, they cut to the core of the question by telling their subjects that they were going to hear stimuli from a continuum. Then they simply asked subjects to try to do what previous research said they couldn't: decide where the stimulus fell on the continuum. If indeed the perceptual speech system is deaf to within category acoustic variation, then the task should have been impossible. Massaro and Cohen, however, reported that subjects could perform this task remarkably well; their ratings did not look like a categorical step function, but rather a smooth continuous line.[9]

Massaro also tackled the McGurk-MacDonald effect. While the effect clearly demonstrates the brain's ability to integrate information arriving from different senses, he questioned the idea that a specialized motor-gesture-based system was the source of the phenomenon. One problem he noticed was that perceptual integration in McGurk-type stimuli was asymmetric. An auditory *ba* paired with a visual *da* more often goes the way of the visual signal: subjects tend to hear *da*. But the reverse pairing, auditory *da* with visual *ba*, usually results in the listener's reporting the auditory stimulus, *da*. This shouldn't happen if the perceptual system is more influenced by the visually perceived *gesture*.

Massaro proposed that we interpret perceptual events in a way that is most consistent with *all* of the available sensory information. In support of this notion, Massaro pointed out that visual and auditory speech cues are largely complementary. For example, /b/ and /d/ sounds are similar acoustically and sometimes confused when the listener must rely only on the auditory signal (it is the reason spelling or radio alphabets were developed, "bravo, delta, foxtrot . . ."), but are easily distinguished visually. Conversely, /p/ and /b/ sounds are visually identical but readily distinguished acoustically. Put the two complementary sources of information together, weight them appropriately for how reliable a cue is, and you get clear speech. Under this view, why do observers typically perceive *da* when presented with an auditory *ba* and visual *ga*? Because acoustically, *ba* and *da* are more confusable than *ba* and *ga*, and visually *ga* and *da* are more confusable than *ga* and *ba*—multiply out the probabilities given all the sensory information and the perceptual system converges on the sound that is most compatible with all the data, *da*. The upshot: the McGurk effect can be explained without reference to a specialized gesture-based module in the brain.[10]

Another blow to the motor theory came when speech scientists took a closer look at whether motor gestures are in fact as invariant as

Liberman suggested. Recall that invariance in the way speech sounds are articulated (as opposed to the resulting acoustics) was a motivating observation behind the development of the motor theory. As it turns out, when you look carefully, the way speech sounds are produced varies substantially.[11] In fact, there is a general problem in motor control referred to as *motor equivalence* that, despite its name, is a lot like the lack of invariance problem in speech perception. Basically, the motor equivalence problem refers to the fact that there are many ways to perform a movement to achieve a particular goal. Think of a golf swing, for example. While there is one goal—to hit the ball—there are nearly endless variations on the swing: head up, head down; straight arm, bent arm; hips open, hips closed; knees bent, knees straight; grip here, grip there. This is a problem for understanding motor control because the number of available options makes it difficult to work out how a movement is planned and coordinated in the brain. For our purposes it raises questions about whether we can turn to the motor system to solve the perceptual invariance problem.

If the motor theory is wrong, how is the lack of invariance problem solved? Massaro proposed a solution. Instead of trying to look for regular acoustic patterns associated with particular phonemes, we should focus on larger-sized units, such as syllables.[12] Just like a small piece of an object, say, the handle on a teacup, is hard to recognize when viewed in isolation but is perfectly obvious when viewed in the context of the whole object, so too bits of the acoustic speech signal seem to be easier to recognize in the context of a whole syllable. In isolation, a small, phoneme-sized unit of speech may sound like a chirp, but in context it forms a crucial part of a very recognizable /d/ sound, for instance.

REBIRTH OF THE MOTOR THEORY OF SPEECH PERCEPTION

BY THE late 1980s, I recall the motor theory of speech perception being presented in graduate school classes as an interesting idea that turned out to be wrong, a sentiment captured bluntly in a review of the theory published in 2006 by a group in the motor theory's birthplace at Haskins labs: "[The motor theory] has few proponents within the field of speech perception, and many authors cite it primarily to offer critical commentary."[13] It's ironic, then, that Liberman's theory provided the theoretical motivation for the Parma group's interpretation of the function of mirror neurons. Neuroscientist and longtime collaborator with the Parma team Marco Iacoboni writes:

> immediately after mirror neurons were discovered in Parma, Giacomo Rizzolatti told Luciano Fadiga that the properties of those neurons reminded him of the motor theory of speech perception of Alvin Liberman (Luciano Fadiga, personal communication).[14]

Luciano Fadiga reiterated his personal communication to Iacoboni publicly at a meeting of the *Society for the Neurobiology of Language* in Chicago in 2010. He and I were participating in a platform debate on the role of mirror neurons in speech perception and as part of his remarks he stated that the motor theory of speech perception was the inspiration for the mirror neuron theory of action understanding. This is reflected also in the first publications on mirror neurons in 1992 and 1996, which featured discussions of Liberman's ideas.

So despite the fact that the motor theory was essentially dead to the community of speech scientists who studied it, the theory enjoyed a resurrection among neuroscientists after the discovery of mirror neurons. Neuroscientists soon took their turn at (re-)examining the empirical viability of the motor theory of speech perception. The interesting (or theoretically unsettling) part was that this fresh look at the motor theory was motivated by the hypothesized function of mirror neurons. To quote one of the early studies authored by Fadiga and colleagues:

> [The motor theory] has been often criticized as too speculative and devoid of firm experimental evidence. More recently, however, the plausibility of the motor theory of speech perception has been reinforced by a series of new experimental data, derived from the neurophysiology of the motor system.[15]

The authors then go on to describe the response properties of mirror neurons and the proposal that they are the brain mechanism for understanding actions. The circularity of the argument is glaring.

The problem, of course (besides the circularity), is that the existence of mirror neurons doesn't erase data that led to the rejection of the motor theory of speech perception a decade before. Infants and chinchillas can still perceive speech quite well without a functional motor speech system, for example. But the old arguments and evidence were ignored, and a steady stream of new studies on the role of the motor system in speech perception has kept journal editors and reviewers busy since the early 2000s. Some of them are impressive and in isolation appear to provide strong support for the motor theory of speech perception. It's worth highlighting a few of the best because the speech domain has become the ultimate test case for the viability of the action understanding interpretation of mirror neurons. Speech research inspired the Parma interpretation of mirror neurons and it has been a foundational case for the generalization of mirror neuron function to humans. If the hypothesis fails for speech, it casts a long shadow on the whole theoretical paradigm of motor simulation.

One of the most highly cited studies was conducted by the Parma group. They tested nine participants using a TMS setup and found that when subjects listened to speech sounds articulated using prominent tongue movements (the double *r* in Italian), listeners' tongue muscles were more twitchy compared to listening to speech sounds that involved minimal tongue movement or to nonspeech sounds.[16]

We can conclude that hearing sounds generated by tongue movements activates tongue muscles in the listener. Does this mean that the neural circuits involved in perceiving the sound of a double r are linked to the motor circuit that can generate that sound? Yes, probably. Does this mean that the double r motor circuit is necessary for perceiving the sound? No. The experiment documents an association, not the causal involvement of the motor system for perceptual recognition. And the speech perception abilities of prelingual infants and chinchillas as discussed previously underline the need for caution in interpreting these effects causally.

Several subsequent studies aimed to try to establish such a causal relation. One of my favorites used TMS to stimulate either the lip or tongue area in the motor cortex of 10 participants while they were asked to identify which of four syllables was presented on each trial. Two of the syllables started with consonants that required prominent lip movement (/p/ or /b/) and two started with consonants that required prominent tongue movement (/t/ or /d/). The syllables were presented against a background of white noise, which dropped baseline performance down to approximately 75 percent accuracy. The logic behind using white noise is that it is easier to see the effects of the experimental manipulation if the stimuli are on the edge of perceptibility. If the task is too easy, a small experimental effect is hard to detect—a so-called ceiling effect. Subjects identified syllables via button press and their choices and reaction times were recorded.

The authors reported that when the lip motor cortex was stimulated, subjects were faster to respond to lip sounds; they also misidentified tongue-related sounds as lip-related sounds more often. The reverse pattern was reported for tongue motor cortex stimulation. Stimulating the motor cortex associated with speech articulators appeared to causally affect the perception of speech sounds in a specific manner. The authors concluded that "these results provide strong support for a specific functional role of motor cortex in the perception of speech sounds."[17]

First let's remember prelingual infants and chinchillas. Any interpretation of the role of the motor system in speech perception must address the whole range of relevant facts, not only an impressive-sounding TMS study or two. So right off the bat we know that whatever the motor system is doing, apparently it is not necessary for speech perception to occur. Now, in this context let's look a little closer at the lip versus tongue stimulation study that at first appears to be an airtight case for the critical involvement of the motor system in speech perception. In fact, it has more than one hole.

Hole number one: it appears that the team didn't just stimulate motor cortex, but it also zapped somatosensory cortex. Many studies that investigate TMS-induced activity in the motor system identify their neural targets directly by moving the TMS device from place to place to find the site that, when stimulated, gives rise to the largest twitch in the target muscle. In the lip and tongue study, the stimulation sites were determined *indirectly* from a previous fMRI experiment of lip and tongue movement. This introduces several sources of error including variability among subjects, fMRI localization accuracy, and the fact that moving your lips and tongue not only activates the motor system but somatosensory regions: when we move, we also *feel* the consequences of that movement.

The team reported the coordinates of the sites they stimulated and indeed, based on standardized functional-anatomic maps, the somatosensory system was likely stimulated. It is hard to conclude that the motor system influenced speech perception when TMS stimulation wasn't accurately targeting the motor system. Further, if the somatosensory stimulation resulted in lip tingling on some trials and tongue tingling on others, we have to worry about whether this might have tipped the hand of the experimenters' hypothesis and led to a bias in the response of the subjects. In the context of barely perceptible stimuli (recall the white noise), clever subjects can and do use any cue to give the response they believe the experimenter is looking for. *Ah, my lips are tingling. It must be a lip-related sound.*

Hole number two: The goal of the TMS study was to assess the role of the motor system in the perception of speech sounds. But the task they used, like many psychological experiments, involved more than just perception. It involved *deciding* among response alternatives on the basis of partial (noisy) information. As we all know, decisions are subject to bias, expectations, and other factors. The question is, did the TMS manipulation affect the perception or the decision process? There *is* a way to tell.

It involves a method called signal detection, an approach developed during World War II to improve detection of enemy aircraft from (sometimes noisy) radar signals. In a nutshell the method compares the proportion of "hits" from an observer (correctly identified signals) with "false alarms" (saying that a signal was present when it wasn't) to calculate detectability with decision bias factored out. A direct measure of bias can also be calculated (basically the probability that a subject decides one way or another, independently of the stimulus). Unfortunately, the authors of the TMS study didn't use these methods. In fact, in the analysis of the participants' accuracy performance the researchers only examined "misses"—when one stimulus was presented and the subject said it was another stimulus—and did not count "hits" at all. This kind of analysis is susceptible to the influence of subject biases.

A 2012 functional imaging study carried out in my lab was designed to uncover the brain correlates of response (decision) bias and substantiates the concern raised above. Subjects were scanned while they performed a task in which they decided whether two phonemically similar syllables were the same. As in the TMS study, the syllables were presented in noise to make the perceptibility harder. Then, without changing the perceptual difficulty of the task, we manipulated the response bias of the subjects by telling them that there would be a greater or a lesser proportion of "same" trials. If you have incomplete perceptual information but you know that two out of three trials will be "same," you tend to respond "same" more often. A bias measure was calculated and, as expected, it showed that the manipulation worked. We then looked at the imaging data and we found that changes in bias were correlated with changes in brain activity in motor and somatosensory speech areas, including the same areas stimulated in the lip-tongue TMS study.[18]

Hole number three involves the task in the TMS study: identifying or discriminating meaningless syllables. A large majority of speech science is founded on such tasks. The logic behind their use is that simple, meaningless syllables allow us to home in on the level of processing we are interested in studying, the perception of speech sounds, while factoring out higher-level nuisances like word *meaning*. Unfortunately, it turned out that by isolating speech sounds and asking listeners to make perceptual decisions about those sounds, researchers ended up studying a partially different cognitive process compared to listening to speech under more natural circumstances.

This point is personal for me so I'm going to spend a few lines on the story. When I was a postdoctoral scholar at MIT, a graduate student in the same department, David Poeppel, started challenging the standard dogma about the left-dominant lateralization pattern of speech perception. Poeppel was throwing around the idea that speech perception is bilaterally organized (involves both the left and right hemispheres) rather than the standard view that perceiving and comprehending speech is the purview of the left hemisphere alone.

The data for his arguments came from two sources. One was the observation that listening to speech resulted in electromagnetically measured responses in both hemispheres. The other was a syndrome called *pure word deafness*. This syndrome is usually caused by two separate strokes in the same unlucky person, one affecting left auditory areas and the other affecting right auditory areas. A profound deficit in speech perception results (patients report that speech sounds like foreign language or like it is just going too fast to decode) but spares basic hearing as measured by a standard hearing test. So the physiological response to speech is bilateral and bilateral damage to speech-auditory areas causes a profound speech perception deficit. Poeppel reasoned that the century-old idea that speech perception is a left hemisphere function is wrong; speech perception is bilaterally organized.[19]

At the time, I thought he was crazy. There was a good reason why the field adopted the view that speech processing is controlled by the left brain. Damage to the left hemisphere results in aphasia, language deficits of a variety of sorts including auditory comprehension deficits, whereas damage to the right hemisphere generally does not.

Further, there was a well-known explanation of the anatomy of pure word deafness that preserved the left dominant view.[20] Other work addressed why the right hemisphere might respond to speech even when it wasn't particularly involved in processing it at the phonological level. A popular idea held that the right hemisphere analyzed prosodic or emotional aspects of the speech signal, for example.[21] All this led me to ignore what Poeppel was trying to say—until I decided to look closely at the evidence supporting the left-dominant doctrine.

A group of undergraduates prompted me to reexamine the evidence. Well, sort of. I was teaching an upper division course in the late 1990s called "Language and the Brain" at UC Irvine and was prepping a lecture on speech perception. One thing I learned about teaching undergraduates is that their questions can be among the most difficult to answer because they are not saddled with the shared assumptions of the field. If, in 1998, I had given a lecture at a conference and mentioned in passing that speech perception is left dominant, no one would have questioned it. Undergrads tend to question those kinds of things, so in my lecture I wanted to show a few slides that presented evidence for the assumption. I went to the literature to locate studies where damage to left superior temporal lobe speech-related areas (classical Wernicke's area) was shown to cause significant speech perception deficits. The evidence didn't exist. In fact, I found evidence that damage to left auditory areas didn't seem to disrupt speech perception at all, but instead caused speech *production* deficits![22] These production deficits were a puzzle to me at the time but turned out to be a critical clue that set my career off in a new direction, one that would ultimately collide with mirror neurons. But let's stick to the perception story.

What I found wasn't conclusive evidence against a left-dominant view. It could be that the relevant regions for speech perception were not in classical Wernicke's area (the posterior superior temporal gyrus) but at a more complex level of auditory analysis involving some other nearby or more widespread left-hemisphere region(s). We know that wider damage to the left temporal lobe can result in Wernicke's aphasia, a syndrome in which the ability to understand spoken words is impaired. I dug into the literature a little deeper and found that the auditory comprehension deficit in Wernicke's aphasia was not caused by a profound speech perception deficit, as was once assumed, but instead seemed to arise from difficulties at the semantic or sentence level of processing.[23] And I knew that damage to motor speech areas didn't cause speech perception problems because patients with those lesions have good auditory comprehension. You need to be able to perceive speech sounds to comprehend words. In short, my literature search revealed that damage to the left hemisphere—anywhere in the left hemisphere—did not cause substantial speech perception deficits. Only bilateral damage to the superior temporal lobe consistently resulted in dramatic perception deficits.

I called Poeppel and told him that he was right after all. He said, "duh," and we decided to write a joint paper that laid out the evidence for a bilateral speech perception system.

Here's where this story relates to the TMS studies and the task they employed in particular. Poeppel and I wrote a modest paper that reviewed the evidence we had collected on the organization of speech perception systems in the brain and submitted it to *Neuron*, a high-impact journal in neuroscience. We got three peer reviews back. One said that it was great and recommended publication. One said that it is common knowledge that speech perception involves both hemispheres and it was too trivial a claim to warrant publication. The third said that our arguments can't possibly be correct because there is abundant evidence that left hemisphere damage causes speech perception deficits. In the end, the journal's editor rejected the paper because it was too controversial for that particular forum. The relevant bit for the story here is the evidence cited by the third reviewer. This evidence showed that damage to many different left-hemisphere regions, but particularly left frontal areas, caused deficits in the ability to label or discriminate meaningless syllables.[24] Poeppel and I had ignored this literature because we were focused on the ability to process speech sounds in the context of a more naturalistic task, that of understanding words.

I went back to the literature to figure out what was going on, and found a paradox. The ability to discriminate similar-sounding pairs of meaningless syllables "doubly-dissociated" from the ability to distinguish these same syllables in the context of a word comprehension task. A double-dissociation is the gold standard for concluding the independence of two processes. To illustrate, imagine an automatic coffee machine that also makes tea. You're curious about whether the same mechanism makes both the coffee and the tea. One day you find that the machine is making tea but won't make coffee. This is a single or one-way dissociation. While it *suggests* separate mechanisms for the two beverages, it isn't conclusive: maybe coffee is just harder to make than tea and the machine developed a partially clogged or weak mechanism that is sufficient only for tea. You get the machine repaired and some weeks later you find that it won't make tea but will make coffee. This, together with the previous observation, is a double-dissociation. Now you can conclude with confidence that somewhere in that machine are (at least partially) different mechanisms for making the two drinks, otherwise how could you lose tea and not coffee *and vice versa?* It can't simply be an effect of difficulty.

Returning to speech, experiments had shown that some patients could not reliably tell whether a pair of syllables such as *ba-da* were the same or different but could understand the difference in meaning between, say, *bad* and *dad*. Other patients showed the reverse dissociation; they could easily discriminate *ba* from *da* but had trouble with understanding the meaning of words.[25] The double-dissociation suggested that the performance of the two tasks—"syllable discrimination," as it is called, and word meaning comprehension—relied on at least partially different mechanisms.

The most puzzling aspect of these findings is that some patients couldn't say whether a pair of syllables was different or the same yet could understand words that contained those same syllables. How is that possible? If an individual can't perform a syllable discrimination task (*ba-da*, same or different?), a natural interpretation is that the speech perception system is failing to process the sounds correctly; there is a fundamental deficit in *hearing* the sounds. But if this were the case, how could this same individual correctly understand words that contain, indeed are differentiated by, those same speech sounds? It had to be that the deficit on the syllable discrimination task did not arise from a deficit in *hearing* the speech, but rather stemmed from a disruption to another mechanism. The *task* of syllable discrimination, Poeppel and I concluded, was calling on some other process besides basic speech perception.

Looking more broadly in the literature, I noticed that discrimination-type tasks tended to involve parietal-frontal circuits, whereas the comprehension tasks involved temporal lobe circuits—a "neuroanatomical double-dissociation." Further, the parietal-frontal circuits overlapped quite well with brain regions that play a role in verbal short-term memory (such as temporarily remembering a phone number), which relies on speech production (the ability to mentally *rehearse* the phone number). Perhaps, we reasoned, syllable discrimination relied on verbal short-term memory in the sense that it required the subject to listen to one syllable, remember it for a short period of time (usually one second), listen to another syllable, and compare the two. This would explain why it is possible to fail on the discrimination task, because of a short-term memory deficit, yet succeed on word comprehension tasks, because there is no requirement to hold two syllables in memory and compare their sounds. It also would explain why deficits in syllable discrimination were associated with parietal-frontal lesions, the dorsal stream, because verbal short-term memory involves speech production systems.

With a nod to the then recent proposal regarding dorsal *how* and ventral *what* streams related to vision, we morphed our review paper in our own "dual stream" model of speech processing, linking speech processing for comprehension to the ventral, temporal lobe stream and the performance of discrimination tasks to the speech production–related dorsal (frontoparietal) stream.[26] It turned out that the *task* matters when designing speech experiments because the task determines which neural mechanisms, or "apps," are used.

Just to hammer home how the task can affect the process and therefore the neural circuits involved, pay close attention to the next sentence. *We do not hear phonemes during normal listening, we "hear" words or more precisely word and phrase meanings.* Now, without looking back, ask yourself, was there a *ba* in the last sentence? Most likely you have no idea. If you tried to figure it out you probably silently "re-spoke" the sentence in your head, thus involving frontoparietal speech circuits, the same circuits implicated in syllable discrimination tasks. Now try this: tell me whether there is a *ba* in the next sentence. *The quick brown fox jumps over the lazy dog.* You can do it but your attention has shifted from comprehending the words to the sound of the phonemes. This process isn't used in normal comprehension and it doesn't come naturally. In fact, the ability to read is quite dependent on the ability to consciously attend to the sound of phonemes, so-called *phonological awareness*, and it is a skill that must be taught explicitly and is difficult for a sizable fraction of individuals to master. Interestingly, there is evidence that illiterate individuals have substantial difficulty performing tasks that require attention to phonemes.[27]

Returning now to the lip and tongue stimulation study, I claimed that the task they used, syllable identification, was one of the holes in their theoretical argument. The task requires conscious attention to the sound structure of speech stimuli, which invokes processes that are not involved in everyday speech analysis. And in fact, empirically, the syllable identification task (hearing a syllable and indicating which of several printed letters matches) has proven to be an even more demanding task in this respect than syllable discrimination, perhaps because it requires consciously attending to the sound and then matching the sound to a written form. For all we know, the TMS could have disrupted the sound-to-letter matching part of the task. If we are interested in understanding the neural basis of the ability to consciously identify or discriminate meaningless syllables, then using such tasks is perfectly valid (and may provide useful information for our understanding of reading). But if we are interested in understanding speech perception as it occurs outside the laboratory, then the results of such studies should be viewed with extreme suspicion.

A handful of additional TMS studies report similar findings beyond the one we've discussed. Stimulate motor-speech areas and you can show (rather modest) effects on "speech perception."[28] But all of them are open to a similar kind of critique: they use partially ambiguous stimuli, fail to account for response bias, and use artificial tasks that don't necessarily tap into processes that are used under normal, every-day speech listening conditions.

TMS isn't the only method employed recently to argue for a neo-motor theory of speech perception. Results of fMRI studies showing that listening to speech activates motor speech areas have also provided fodder for motor accounts of perception.[29] fMRI, however, is purely correlative. Yes, fMRI shows definitively that motor speech areas activate during speech listening. However, the method cannot tell us what function the activated areas serve, nor can it tell us whether such areas are causally related to the process of interest.

LANGUAGE DISORDERS ILLUMINATE THE DEBATE

DECADES OF research on the motor theory of speech perception ultimately led to the theory's rejection. But the discovery of mirror neurons, whose interpretation was inspired by the (dead) theory, led several teams to reinvestigate the question using TMS and functional imaging studies. The TMS studies use tasks of dubious relevance to normal receptive speech function and the fMRI studies are inconclusive with respect to the functional relevance of the activation patterns. Such findings, in the context of the data from chinchillas and infants, do not justify the motor theory's resurrection. Nonetheless, the theory lives, in zombie form I would argue, as the titles of some recent papers attest:

- "The Motor Somatotopy of Speech Perception" [*somatotopy* means "body map"][30]
- "Motor Representations of Articulators Contribute to Categorical Perception of Speech Sounds"[31]
- "The Essential Role of Premotor Cortex in Speech Perception"[32]

The renewed popularity of the motor theory motivated my colleagues and me to review the literature and kick off a new round of experiments aimed at reexamining the role of the motor system in speech perception. Here is some of what we found.

If the motor speech system is critical for speech perception, we should expect neurological disorders that prevent the ability to speak to produce a measurable decline in the ability to understand speech. A seminal case study on the topic was reported in 1962 by Eric Lenneberg, a German-born scholar who fled his home country as a young man to escape Nazi persecution. In the 1962 report, Lenneberg describes his observations of the medical history and language ability of a young boy whom he studied for more than four years.[33] The boy was brought to the neurological service and to Lenneberg's attention because at the age of three years and nine months he had failed to develop speech. All laboratory tests were normal, he had no anatomical abnormalities in his oral cavity, and he could chew, swallow, suck, blow, and lick normally. Medical records indicated that he was born with club feet and at approximately age two, strabismus (misalignment of the eyes) was noted and then corrected surgically. IQ tests revealed scores around the lower boundary cutoff for normal, but it was unclear how much of this "deficit" could be attributed to his inability to express himself verbally.

Lenneberg noted that the child's ability to speak was effectively nil. The child cried and laughed normally but made only cough-like grunts that accompanied his attempts to communicate via pantomime. When he played alone, he often made sounds that resembled Swiss yodeling, although he had never been exposed to the art form. Lenneberg further noted that as the child got older, and after years of speech therapy, he could repeat a few words after his therapist or mother, but even these attempts were "barely intelligible" and were never produced on his own.

Lenneberg was fascinated by the boy's ability to comprehend language, which the scientist characterized as "normal and adequate understanding of spoken language." This observation was confirmed over more than 20 visits and by several doctors and speech therapists,

both informally and formally. The boy could follow complex commands such as "Take the block and put it on the bottle" and he could
appropriately respond, albeit nonverbally, to questions about a short
story that was read to him. He was not merely cuing off the body
language of the researchers: he accurately followed commands even
when he couldn't see the investigator. The cause of the speech production deficit could not be determined. Lenneberg called it "congenital
dysarthria"—difficulty controlling the speech articulators—and concluded that the ability to master receptive language, to understand,
did not depend on the ability to produce or imitate speech.

Today we know that a similar symptom pattern in children,
Foix-Chavany-Marie syndrome, can be caused by bilateral lesions involving a brain region called the *anterior operculum*.[34] This area is close to
Broca's region and has been implicated in motor speech function. The
disorder is considered a form of paralysis of the motor pathways that
control the mouth and facial muscles (although it only affects voluntary
control, not involuntary control). As such one might write it off as
irrelevant to the neuroscience of higher-level speech-language function. This is a fair concern because one might not expect a lower-level
nerve problem to affect more complex linguistic processes in the cerebral cortex. However, we must again wonder whether the use-it-or-
lose-it principle might come into play: does the inability to control
lower-level aspects of speech impact the functioning of higher-level
motor speech centers that are at the center of the mirror neuron debate?

Similar arguments regarding the dissociation between receptive
and expressive speech abilities come from a study of young people with cerebral palsy by developmental neuropsychologist Doro-
thy Bishop, now at Oxford University.[35] Cerebral palsy is caused by
injury or abnormalities in the developing brain and can result in a
variable mix of deficits and severities. Bishop and colleagues studied
two groups of affected young people, one in which severe deficits
in speech production were present and another, matched to the first
on IQ and age, with normal speech production. The study authors
reported the results of one experiment that at first seems to support
the motor theory perspective. The speech-impaired group performed

significantly worse than the normal speech group on a test of syllable discrimination. This was true even when measures were used that correct for response bias. I consider this fairly solid evidence that motor speech dysfunction affects the ability to discriminate meaningless pairs of syllables.

However, again that pesky task issue arises. Is syllable discrimination the right test? Is it just tapping into verbal short-term memory rather than speech perception per se? Bishop and her team carried out a second experiment to find out. The group devised a clever variant of the task that they called a word judgment task. Here's how they described it to their young participants (in slightly paraphrased form):

> I'm going to show you some pictures. For each picture I will say the name of what is in the picture, but sometimes I will say it a bit wrong. For instance, I might show you this picture (a boy) and instead of saying, "Is it a BOY," I say, "Is it a VOY?" You have to say if I am right or wrong.

So the task still requires the listener to detect the difference between similar-sounding phonemes, but the presentation is more naturalistic. To perform the task you don't have to consciously listen to the sounds themselves, you only have to take note of whether the syllable corresponds to the picture. A standard syllable discrimination test was also given using the same set of phoneme contrasts (e.g., /b/ vs. /v/, among others). On the standard syllable discrimination task, the speech-impaired group again performed significantly worse compared to the normal-speech group. However, on the word judgment task, the scores of the speech-impaired group rose substantially and the two groups did not differ in performance.

There are two, now familiar, take-home messages from this study. Number one: the inability to produce speech does not preclude the ability to perceive even subtle differences between speech sounds. This was clearly shown by the word judgment task. Number two: standard syllable discrimination tasks do not measure the full capacity of a listener's ability to perceive speech sounds.

To me these studies argue convincingly against the motor theory of speech perception. But zombies are hard to kill, so in my own lab we decided to try to uncover additional relevant evidence.

In one set of studies we revisited the relation between damage to Broca's area—the core of the human mirror system—and the ability to perceive speech. If the mirror neuron/motor theory is correct, damage to Broca's area should impair speech perception. For one experiment I teamed with Italian neurologist and aphasiologist Gabrieli Miceli. Miceli was in the trenches during the 1970s and 1980s push to understand the neural basis of speech perception and auditory comprehension deficits in aphasia. He published more than one seminal paper on the topic. I was interested in having a fresh look at some of those original data on syllable discrimination in Broca's aphasics, specifically to see if the deficits might be accounted for by response bias. I sent Gabrieli an e-mail asking if he still had the data and whether he might be interested in collaborating on a reanalysis. He replied that he was indeed interested but the data no longer existed. However, he told me that he had a new dataset that included all the relevant behavioral measures, plus MRI or CT documented lesion information, and he would be happy to collaborate.

We decided to focus on the crux of the issue: does damage to Broca's area cause deficits in speech perception? Miceli's team in Italy set about identifying the subset of patients in his database that had substantial damage to Broca's area while I analyzed the behavioral data gathered from two critical tasks: auditory syllable discrimination and auditory word comprehension. In the latter, patients saw two pictures on a computer screen, say a bear and a pear, and then listened to a word that named one or the other picture. The task was to point to the picture that matched the word. Crucially the names of the two pictures were phonologically similar (*bear/pear*), which required subtle speech sound perception to distinguish, but without having to consciously attend to the phonemes themselves.

Miceli's group identified 24 stroke patients with lesions involving Broca's area (as well as other regions). These were compared to a control group of 13 stroke patients with damage to the left temporal-occipital regions, that is, a part of the brain not expected to cause language deficits. Nineteen of the 24 cases with Broca's area lesions had been clinically classified as Broca's aphasics, which means their speech production was nonfluent (halting, effortful speech) but their comprehension relatively spared. The severity of nonfluency can and does vary in aphasia and so Miceli's team categorized them into mild, moderate, and severe nonfluent cases. There were roughly equal numbers of each category.

When I looked at the data, using response bias correction analysis methods, I was a bit surprised. Previous reports indicated that patients with Broca's aphasia have deficits on syllable discrimination, but Miceli's data showed remarkably good performance, averaging around 95 percent accuracy even without the bias correction. However, this level of performance was still significantly worse than the control group of nonaphasic stroke patients, who performed virtually flawlessly (it's a very easy task). It was possible that motor speech deficits were indeed impacting speech perception, if minutely.

A closer look suggested otherwise. For example, the severity of nonfluency had no effect on syllable discrimination performance; if motor speech deficits cause perceptual deficits there should be a relation. Severe nonfluent patients should have more trouble with the perceptual task. Further, similar to the study of cerebral palsied children, when we looked at the word comprehension task, performance in the group with Broca's area lesions came up to the level of the controls (uncorrected scores of 97 percent and 96 percent, respectively) even though the comprehension task required fine differentiation of speech sounds. Conclusion: Broca's area damage and deficits in speech fluency do not affect the capacity to differentiate subtle differences in speech sounds.

A larger-scale lesion study spearheaded by a former doctoral student of mine and current collaborator, Corianne Rogalsky, confirmed these results. Fifty-eight individuals with left-hemisphere lesions were studied on a word comprehension task. Words such as *bear* were presented acoustically and the participant was asked to point to the picture on a card that matched the word. The card had four pictures, the one that matched the word, one that was similar phonemically (e.g., pear), one that was similar in meaning (e.g., moose), and one that was completely unrelated (e.g., grapes). This task is very easy, even for stroke patients, so we added noise to the word stimuli to ramp up the perceptual difficulty. Rogalsky then examined the lesion data using an algorithm that picks out regions that, when damaged, result in deficits on the task. The region that was most strongly predictive of auditory comprehension deficits was in the temporal lobe, part of the ventral stream, not Broca's area or parietal regions.[36]

We looked at whether the motor system contributes to speech perception from one final methodological perspective that I'd like to mention. For this one, my lab teamed up with neurologist and epileptologist Arthur Grant to assess the speech perception ability of the left and right hemispheres separately. The Wada procedure is named after neurologist Juhn Wada who developed the technique in postwar Japan in the late 1940s. Here's how it works. The patient lies on a table while a catheter is fed into either the left or right internal carotid artery. The carotid arteries are the blood supply conduits to the brain; one supplies the left hemisphere and the other supplies the right. Once the catheter is safely inside one of the arteries, doctors inject sodium amobarbital, a barbiturate, through the tube, putting one hemisphere to sleep instantly while leaving the other one awake. With one hemisphere functionally offline, the lone capabilities of the other one can be assessed. Once the drug wears off, the procedure is then repeated with the other hemisphere anesthetized to get a complete picture of each hemisphere's functional capabilities.

The Wada procedure is not done simply for research purposes, but rather as a clinical procedure as part of a presurgical workup for neurosurgery. Its clinical utility comes from the procedure's ability to localize speech and memory functions. In most people, the ability to control speech production is localized to left-hemisphere networks while basic memory functions can be achieved by either hemisphere. But functional organization varies from person to person and neurosurgeons like to know whether the brain they are about to cut into is organized typically or atypically.

In our study,[37] we capitalized on the fact that putting the left hemisphere to sleep causes (in most people) a complete loss of the ability to speak. With the left hemisphere anesthetized, the patients are completely mute (as well as paralyzed on the right side of their body). We wondered whether they would be able to understand speech. We knew from previous research that they would, at least to some degree, because patients with their left hemisphere asleep can follow simple commands such as "point to the ceiling." But we wanted to know how well they would perceive subtle differences in speech sounds while comprehending words.

For each of 20 patients, after the clinical assessment, we administered the same auditory comprehension task used in the stroke patients: hear a word (but not in noise) and point to the correct picture from an array of four. If the motor system is critical for perceiving speech sounds, we should expect that completely deactivating the speech motor system should cause substantial deficits in perceiving speech. In the extreme we might expect complete failure to understand the words at all. At the very least, we should see more pointing errors involving the phonemically similar picture, because this would be the most difficult to differentiate from the target if speech perception is compromised.

Despite the fact that all 20 patients were completely mute at the time of testing, they correctly pointed to the matching picture on 77 percent of the trials. This is well above chance level performance (25 percent)—they correctly comprehended a majority of words—but it still represents an impairment relative to the same patients' baseline performance on the task, which was 98 percent. But what kind of errors were they making? They rarely chose the unrelated picture (less than 2 percent of the time) and only occasionally chose the phonemically similar foil picture (7.5 percent of all responses). Where they struggled the most was in differentiating the target from semantically related pictures (the moose, for example). They selected the semantic distractor on 17 percent of all responses, which comprised the majority of all errors (75 percent).

What can we conclude? Complete functional deactivation of the motor speech system does not substantially disrupt the ability to process phonemic information in a word comprehension task. Yes, phonemic errors were greater with left-hemisphere anesthesia compared to baseline or right-hemisphere anesthesia (the latter resulted in errors on less than 5 percent of trials), but the phonemic error rate with the left hemisphere asleep was quite low. *And remember,* not only does left-hemisphere anesthesia deactivate motor systems, it also deactivates auditory-related language systems. The smattering of phonemic errors that we observed may have resulted not from motor system deactivation but from anesthetizing left-hemisphere temporal lobe systems, as the lesion data from Rogalsky's study would predict.

No matter whether you look at it from the perspective of Wada procedures, stroke, cerebral palsy, congenital dysarthria, prelingual infants, or chinchillas, when you take the motor speech system out of the picture the ability to perceive speech sounds remains unaffected. The motor theory of speech perception is (still) wrong. And if the motor theory was the inspiration and foundation of the mirror neuron theory of action understanding, then the latter has serious problems.

BEYOND SPEECH SOUNDS

WHILE THE debate raged over the role of the motor system in the perception of speech *sounds*, neuroscientists and psychologists were beginning to explore a deeper possible role for the motor system in the *semantics*, the meaning of language. The idea is simple and sounds a lot like the mirror neuron theory of action understanding: we understand the meaning of action *words* by relating them to the motor programs for performing the actions themselves. The idea actually predates the discovery of mirror neurons and is part of a larger movement in psychology called *embodied cognition*. This is the notion that our knowledge of the world is grounded directly in our sensory and motor experiences. It contrasts with a "classical" view of cognition, which is often viewed as embracing a more abstract, symbolic view of knowledge and information processing; something closer to how a digital computer codes and processes information. What's at stake in the debate, then, is well beyond action understanding and mirror neurons. It's our understanding of the human mind itself. Mirror neuron theory has both fueled and been fueled by this embodied cognition movement. So much so, in fact, that mirror neurons and embodied cognition have become deeply enmeshed. To understand the relation between the two ideas and the empirical support behind each, it is worth spending some time untangling the theories. To do this we again need to take a step back, theoretically and historically, to put the current debates in their proper context.

6

The Embodied Brain

FOR MUCH of the first half of the twentieth century American psychologists—most prominently John Watson and B.F. Skinner —viewed the mind/brain as a simple association device that links stimulus events in the environment with motor responses. Fire burns hand → hand withdraws; rock hurtles toward head → head ducks; lion roars → organism runs; stimulus → response. Such behaviors are relatively reflexive, but according to these theorists complex behaviors are no different in terms of their underlying associative mechanism. The only difference is that complex behaviors reflect more complex associations, shaped by two simple learning rules:

> *Classical conditioning*–pair one stimulus with another
> stimulus that elicits a natural reflex and pretty soon the
> first stimulus elicits the response all by itself. If sensory
> stimulation from tasting chocolate triggers a natural
> reflex causing your mouth to water, pretty soon the
> sight of chocolate alone causes your mouth to water.
> And if the sight of chocolate is frequently paired with
> seeing your favorite candy store, then the candy store
> alone—or the road that it's on or the car you were in—
> causes your mouth to water. The complexity of behav-

ior is just associations built on associations built on a few innate reflexes.

Operant conditioning–reinforce a particular response to a particular stimulus and the future likelihood of that response given that stimulus increases. This is just the standard animal training principle. If you say "sit" and your dog sits by accident or force, give him some bacon. Next time you say "sit," he is more likely to sit.

The mind, then, held little sway according to this psychology, known as behaviorism. There was no room for beliefs, desires, or thoughts—only associations. In fact, behaviorists believed that because we can't *see* mental states or processes we should not venture to study them, lest we be led astray. Skinner wrote:

When what a person does is attributed to what is going on inside him, investigation is brought to an end. . . . For twenty five hundred years people have been preoccupied with feelings and mental life, but only recently has any interest been shown in a more precise analysis of the role of the environment. Ignorance of that role led in the first place to mental fictions, and it has been perpetuated by the explanatory practices to which they gave rise.[1]

This theoretical landscape began to change in the late 1950s. A new generation of scientists including George Miller, Donald Broadbent, and a young linguist named Noam Chomsky provided clear evidence that human behavior was too complex for simple association mechanisms to explain. For instance, as a freshly minted PhD, Chomsky reacted incredulously to Skinner's 1957 magnum opus, *Verbal Behavior*, which attempted to derive human language capacity from behaviorist principles. We utter the words and sentences that we do in a given context, argued Skinner, because those words and sentences, when spoken previously under similar stimulus conditions, had been reinforced. If we spew the word *Dutch* when viewing a painting by Vermeer, this is only because on previous museum visits we stumbled across the same word while looking at a Vermeer and someone said, "Correct," flashed an approving smile, or handed over a piece of bacon, thus reinforcing the behavior and increasing the probability of saying the same word in a similar situation. The response "Dutch," Skinner would argue, is thus said to be under *stimulus control* by the Vermeer painting. Chomsky's response:

> Suppose instead of saying *Dutch* we had said *Clashes with the wallpaper, I thought you liked abstract work, Never saw it before, Tilted, Hanging too low, Beautiful, Hideous, Remember our camping trip last summer?*, or whatever else might come into our minds when looking at the picture. . . . Skinner could only say that each of these responses is under the control of some other stimulus property of the physical object. . . . This [theoretical] device is as simple as it is empty. Since properties are free for the asking . . . we can account for a wide class of responses in terms of Skinnerian functional analysis by identifying the "controlling stimuli." But the word "stimulus" has lost all objectivity in this usage. Stimuli are no longer part of the outside physical world; they are driven back into the organism. We identify the stimulus when we hear the response. It is clear from such examples, which abound, that the talk of "stimulus control" simply disguises a complete retreat to mentalistic psychology.[2]

This snippet from Chomsky's review conveys his two-part argument loud and clear. Part one: language is too complex for simple stimulus-response psychology to explain. Instead, we need to look inside the mind. Indeed, Chomsky's own work in linguistics was showing how mental processes could explain language behavior: language, he argued, is a *computational* system that takes abstract representations as input (categories of information like NOUNS and VERBS), computes relations between the category representations, and derives a combinatorial whole that is more than a simple sum of its parts.

For instance, *He showed her the baby pictures* and *He showed her baby the pictures* have exactly the same parts but differ in how they are *combined* by the brain's computational system, which encodes different meanings. Further, the system is *productive*: given a basic set of categories and computational rules, it can generate an endless stream of novel combinations (*He showed her the baby pictures, then the cat pictures, then the petunia pictures . . .*). Today we call this approach to cognition the *computational theory of mind*, the idea that the brain is an information processing device that performs computations[3]—more on this idea momentarily.

Part two of Chomsky's argument: relabeling a complex mental function using simple-sounding vocabulary doesn't eliminate the complexity, it only disguises it. The result of Chomsky's critique, his research in linguistics, and the work of his contemporaries in other domains such as working memory, attention, and problem solving was the realization that the mind is not a simple stimulus-response associator but rather an active information processing device.

COMPUTATION AND THE INFORMATION
PROCESSING APPROACH

THE QUESTION then became, how does the brain process information? The digital computer was being developed around the same time and served as a convenient heuristic to think about how the brain might achieve such a feat. The basic idea is that there is information on the one hand and a set of processing routines (mental apps or computational algorithms) on the other. The information serves as input to the processing routines, which then transform it according to the set of computations defined in the program (e.g., if x, then y) and the processing routines output the results of the transformations. The output can then be stored as new information, serve as input to other programs, and, in the case of digital computers, control devices like a display or printer. Inputs to a computer—key presses, mouse jiggles and pokes, image captures—don't directly control what's displayed on the monitor or what the printer prints; rather, those inputs are *processed* by various apps to convert them into words and images, solve math problems, or play solitaire. It is the output of the *apps* that directly controls what is displayed or printed.

The point of the computer analogy, or more accurately the computer *program* analogy, is that the mind/brain works the same way: the inputs to the brain—photons hitting the retina, air pressure fluctuations impinging on the ear drum, and so on—don't control human behavior directly; rather, those inputs are *processed* by various neural apps to convert them to words and images, solve math problems, or play solitaire.

Some early cognitive models looked very much like computer programs. In fact, some *were* computer programs. One of the most famous programs, developed in the 1950s by Allen Newell, J. C. Shaw, and Herbert Simon, was called the Logic Theorist. The Logic Theorist, or LT as it was nicknamed, was written to prove theorems using symbolic logic, similar in spirit to what a high schooler encounters in geometry class. The program was given a database of axioms (e.g., symbolic logic statements like "p or q implies q or p") and a set of processing rules for using the axioms to generate proofs, rules such as *substitution* or *replacement*. LT was then presented with a series of new logic expressions and instructed to discover the proof for each using the "given" axioms and the rules. If it proved a theorem, it could store that proof along with the given axioms for use in subsequent proofs.

LT performed quite respectably, proving 73 percent of the theorems it was given. Writing a program that could pass a high school geometry class was an impressive accomplishment for the infant field of computer science, but it had far more significance for the information processing approach to understanding human behavior. LT showed that complex, human-like behavior could be approximated quite well with a purpose-built information processing system. And this is how Newell et al. presented the LT program, as a straight-up theory of human problem solving. To bolster their argument, the team presented evidence that LT's problem-solving "behavior" exhibited features characteristic of humans' solving similar problems, such as its ability to learn, its demonstration of a kind of "insight" (trying at first to solve a problem with trial and error and then, once hitting upon the solution, using the same approach to solve similar problems), and its ability to break a problem down into subproblems.

Now, Newell and company were careful to point out that their theory does not imply that humans are digital computers, only that humans appear to be running a program similar to LT:

> We wish to emphasize that we are not using the computer as a crude analogy to human behavior—we are not comparing computer structures with brains, nor electrical relays with synapses. Our position is that the appropriate way to describe a piece of problem-solving behavior is in terms of a program: a specification of what the organism will do under varying environmental circumstances in terms of certain elementary information processes it is capable of performing. This assertion has nothing to do—directly—with computers. Such programs could be written (now that we have discovered how to do it) if computers had never existed. A program is no more, and no less, an analogy to the behavior of an organism than is a differential equation to the behavior of the electrical circuit it describes. Digital computers come into the picture only because they can, by appropriate programming, be induced to execute the same sequences of information processes that humans execute when they are solving problems. Hence, as we shall see, these programs describe both human and machine problem solving at the level of information processes.[4]

The idea that the brain executes programs is not as crazy as it might at first sound. Consider a computer program that is designed to tell you the location from which sounds are coming. Imagine that input comes from two microphones, one on the left and one on the right of a laptop or smartphone, positioned several centimeters apart. Because sound travels at a known and relatively slow speed (approximately one foot per millisecond), if the sound source is off to one side relative to the two microphones, a small but measurable difference occurs in the arrival time of the sound at the two speakers. With this information you can easily write a program to calculate the location of the sound source relative to the orientation of the microphones. A bare-bones program might look something like this:

> Let x = time of sound onset detected at left speaker
> Let y = time of sound onset detected at right speaker
> If $x = y$, then write "straight ahead"
> If $x < y$, then write "left of center"
> If $x > y$, then write "right of center"

If you wanted to get fancy you could calculate the precise relation between x and y and determine exactly how far from the center a sound is and maybe you could even use this information to drive a motor that moved a camera that would point to the location to "see" the source of the sound.

Humans and many other animals can localize sounds in space and it is generally agreed that the brain can calculate location based on the difference in arrival time of a sound to each ear, so-called *interaural time difference* (other cues such as loudness difference can also be used). But you can't write lines of code into brain tissue. Surely, then, the brain isn't implementing anything like this kind of program, right? Well, consider how the barn owl brain does it, a system that has been studied extensively and on which there is good agreement (the mammal system is still being debated).[5]

Inputs from the two ears converge in a particular

structure within its brainstem, the *nucleus laminaris*, but they don't converge randomly. Instead the network of cells where the inputs converge are arrayed into sequential lines, something like in the figure below:

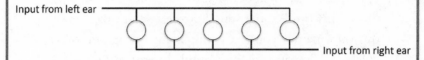

Input from left ear

Input from right ear

Given this arrangement, the neurons (circles) on which the left- and right-ear signals converge *simultaneously* depend on the time difference between excitation of the two ears. If both ears are stimulated simultaneously (sound directly in front), convergence happens in the middle of the *delay line*, as these neural arrangements are called. If the sound stimulates the left ear first, convergence happens farther to the right in this schematic (left-ear stimulation arrives sooner, allowing its signal to get further down the line before meeting the right-ear signal). And vice versa if right-ear stimulation arrives sooner. This delay line architecture basically sets up an array of coincidence detectors in which the position of the cell that detects the coincidence *represents information*: the difference in stimulation time at the two ears and therefore the location of the sound source. Then all you have to do is plug the output (firing pattern) of the various cells in the array into a motor circuit for controlling head movements and you have a neural network for detecting sound source location and orienting toward the source.

Although there is no code in the barn owl's brain, the architecture of the network indeed implements the program: x and y are the input signals (axonal connections) from the left and right ears; the relation between x and y is *computed* via delay line coincidence detectors; and "rules" for generating an appropriate output are realized by the connections between various cells in the array and the motor system (in our example). Brains and lines of code can indeed implement the same computational program. Lines of code do it with a particular arrangement of symbols and rules; brains do it with particular arrangement of connections between neurons that *code* or *represent* information.

It was a one-two punch for behaviorism. Chomsky and others had pointed out the inadequacy of simple associationist explanations of human behavior and symbol-manipulating computer programs provided a viable and directly observable account of how the brain processes information. Psychology wholeheartedly embraced the information processing approach.

The movement was later termed the *cognitive revolution*, an unfortunate term in my view because it implies that the revolution holds only for the everyday definition of "cognitive," higher-order functions like language, memory, problem solving, and the like. It is true that the majority of the earliest work in the field focused on these complex human behaviors, but the real point of the revolution was that everything about human behavior—perception, motor control, all of psychology—is a result of information processing. Psychologist Ulric Neisser, who literally named the field and wrote the book on it with his 1967 text, *Cognitive Psychology*, defined the domain of cognition this way:

> "Cognition" refers to all the processes by which the sensory input is transformed, reduced, elaborated, stored, recovered, and used. . . . Such terms as *sensation, perception, imagery, retention, recall, problem-solving,* and *thinking,* among many others, refer to hypothetical stages or aspects of cognition.[6]

Neisser's table of contents underlined his view that cognition was not limited to higher-order functions. His volume is organized into four parts. Part I is simply the introductory chapter. Part II is called "Visual Cognition" and contains five chapters. Part III is "Auditory Cognition" with four chapters. Finally, Part IV deals with "The Higher Mental Processes" and contains a single chapter, which Neisser refers to as "essentially an epilogue" with a discussion that is "quite tentative." He continues:

> Nevertheless, the reader of a book called *Cognitive Psychology* has a right to expect some discussion of thinking, concept-formation, remembering, problem-solving, and the like. . . . If they take up only a tenth of these pages, it is because I believe there is still relatively little to say about them. (pp. 10–11)

Most scientists today working on perception or motor control, even at fairly low levels, would count their work as squarely within the information processing model of the mind/brain and therefore within Neisser's definition of cognition. Consider this paper title, which appeared recently in a top-tier neuroscience journal: "Eye Smarter than Scientists Believed: Neural Computations in Circuits of the Retina."[7] If anything in the brain is a passive recording device (like a camera) or a simple filter (like polarized sunglasses) it's the retina, or so we thought. Here's how the authors put it:

> Whereas the conventional wisdom treats the eye as a simple prefilter for visual images, it now appears that the retina solves a diverse set of specific tasks and provides the results explicitly to downstream brain areas. (p. 150)

"Solves a diverse set of specific tasks and provides the results" sounds like a purpose-built bit of programing—in the retina! We observe similar complexity in the control of simple movements, such as tracking an object with the eyes, an ability that is thought to involve a cerebral cortex-cerebellar network including more than a half dozen computational nodes that generate predictions, detect errors, calculate correction signals, and learn.[8] It is not much of an overstatement to say that there is universal agreement among perceptual and motor scientists in neuroscience and psychology that perception and action are complex systems that actively transform sensory information and dynamically control action. As Neisser wrote in 1967, "*Information* is what is transformed, and the structured pattern of its transformations is what we want to understand."[9] The information processing model of the mind—cognitive psychology as defined eloquently by Ulric Neisser—now dominates the study of the mind/brain from computation in the retina to motor control to complex problem solving.

EMBODIED COGNITION: AN ALTERNATIVE TO INFORMATION PROCESSING?

NOW, FINALLY, we are in position to appreciate how embodied cognition—also called *grounded* or *situated cognition*—as well as mirror neuron theory fit within cognitive psychology more broadly. Embodied cognition argues in essence that all of cognition comes down to sensory and motor systems (including "internal sensing" such as emotions) and is frequently touted as a radical departure from the classical view of cognition and the mind prompted by Neisser, Chomsky, and their contemporaries. Some researchers are even hinting at a possible paradigm shift[10] in the study of the mind/brain and refer to the movement as *postcognitivism*.[11] But perplexingly the "classical" view that embodied cognition theorists are shifting from is not the cognitive psychology of Neisser, but a different view of the mind in which cognition is distinct from, and sandwiched between, sensory and motor processes—the so-called classical sandwich:

> The classical sandwich conception of the mind—widespread across philosophy and empirical sciences of the mind—regards perception as input from world to mind, action as output from mind to world, and cognition as sandwiched between.[12]

As an alternative to this supposed classical sandwich view, embodied cognition theorists promote a model in which (high-level) cognition is embedded in sensory and motor systems:

> The first and foremost challenge is that cognition cannot be studied as a module independent from other modules (sensory and motor), as suggested by the "cognitive sandwich" metaphor. Instead, cognition is deeply interrelated with sensorimotor action and affect.[13]

> These [embodied] approaches conceive of cognition and behavior in terms of the dynamical interaction (coupling) of an embodied system that is embedded into the surrounding environment. As a result of their embodied–embedded nature, cognition and behavior cannot be accounted for without taking into account the perceptual and motor apparatus that facilitates the agent's dealing with the external world in the first place, and to try to do so amounts to taking this external world also into account. This tells directly against the aforementioned cognitive sandwich.[14]

We have a terminological confusion here. The revolution of the 1950s crashed through the behaviorists' roadblocks to studying the internal organization of the mind and argued instead that all of mental function—sensing, acting, communicating, thinking—is the result of information processing. It became known (unfortunately) as the *cognitive* revolution. Decades later, embodied cognition theorists define cognition, and its "classical" revolution in psychology, as aligned with the term's colloquial definition, applying to higher-level mental functions and excluding sensory and motor systems. The "classical sandwich conception of the mind" is therefore a straw man and an easy target for researchers who endeavor to show that cognition is not distinct from sensory and motor systems.

In this context the embodied cognition movement isn't a "post-cognitive" (or more accurately post–information processing) revolution at all: If seeing, hearing, feeling, and acting involve information processing and therefore involve cognition in Neisser's sense (a relatively noncontroversial assertion), then "grounding" cognition in sensory and motor systems amounts to grounding cognition in cognition.

Once you start looking inside the brain you can't escape the fact that it processes information. You don't even have to look beyond a single neuron. A neuron receives input signals from thousands of other neurons, some excitatory, some inhibitory, some more vigorous than others. The output of the neuron is not a copy of its inputs. Instead its output reflects a weighted integration of its inputs. It is performing a *transformation* of the neural signals it receives. Neurons *compute*. This is information processing and it is happening in every single neuron and in every neural process whether sensory, motor, or "cognitive."

Now some theorists want to push cognition even further out of the brain and into the world, as the last of the previous quotes hints—a kind of grounding in the environment. Here's a more direct comment on this issue from the same authors:

> Illustrations of insightful applications within post-cognitivism include using pen and paper, counting with one's fingers, and drawing Venn diagrams. These actions permit us to off-load cognitive cargo into the world.[15]

No one doubts that the physical properties of the world and the body matter for brain function. After all, it is precisely this environment that the brain evolved to deal with and so *of course* the form of mental operations is influenced by the world and body and takes advantage of what is available therein. But it is equally important to recognize that we can't push the explanation of behavior all the way out of the brain. Been there, done that—behaviorism doesn't work. And to use the environment to "off-load cognition" as illustrated above, you have to have the mental chops in the first place to invent writing, enumerate and recognize that the fingers on your hand can represent numbers, group information into sets, and graphically represent relations between sets. There's a reason why chimps don't write, count on their fingers, or trace Venn diagrams in the dirt.

CONCEPTUAL KNOWLEDGE: A SANDWICH IN THE MIND

WHERE DID the "classical sandwich conception of the mind" come from? Is it merely a terminological misunderstanding? If we look closely at some of the domains of cognition in which embodied theories are gaining traction, answers begin to emerge. One domain, for example, concerns the question of how the brain represents conceptual knowledge, the meanings of objects, events, and actions. What we will see is that embodied cognition is an alternative to a classical theory of conceptual knowledge, not an alternative to the classical theory of the mind. Research on conceptual knowledge attempts to understand how, for example, we come to believe that a hammer and a screwdriver are members of the same conceptual category (TOOLS) and are therefore similar in several respects (man-made, used for fixing or building things, found in toolboxes and hardware stores, and so on). In this domain there *is* a classical model, rooted in philosophy, in which conceptual knowledge is mentally represented in abstract form quite distinct from the sensory and motor systems involved in perceiving and using hammers and screwdrivers.[16] According to this classical view, if someone wanted to hammer a nail and she saw a hammer in a drawer, the visual image would be processed in the visual system, the output of the visual analysis would serve as the input to the conceptual ("cognitive") system that would recognize it as a hammer. From there the conceptual system would relay this information, and the intention to use the object, to the motor system that would implement a command to hammer the nail.[17] This classical proposal looks a lot like a cognitive sandwich.

An alternative to this approach began to emerge from research in the 1970s and 1980s. One key piece of evidence came from the discovery that brain damage could impair conceptual knowledge in one category (e.g., TOOLS) while leaving another category (e.g.,

ANIMALS) relatively unaffected[18] and further that the brain regions involved in causing these deficits included sensory and motor areas.[19] This suggested that conceptual knowledge is not completely sandwiched between sensory and motor systems but instead is somehow tied into those circuits.

Observations such as this eventually morphed into the notion of an embodied model of conceptual representation. To summarize the basic idea, your knowledge of hammers, what they look like, what their function is, and how you use them is not a *high-level* cognitive abstraction stored in a neural database distinct from sensorimotor systems, but rather is the sum total of your *lower-level* sensory and motor experiences with hammers. Like a reporter in a war zone, concepts are embedded within the sensorimotor operations themselves. And if concepts are fundamentally sensorimotor, then "cognitive" operations involving those concepts (categorizing, making inferences, recall) are nothing more than reenactments of sensorimotor experiences. When you think about hammers, the theory goes, you *simulate* your sensory and motor experiences with hammers; a kind of physical thinking as opposed to abstract, logical thinking. In short, concepts are embodied and cognition is reduced to simulation—classical sandwich disassembled.

Images in your head

There is reasonably good evidence that the brain simulates experience on some level. Imagine walking from the front door of your home to your bedroom. How many windows are visible on your walk? Imagine kicking a soccer ball as hard as you can. Where on your foot do you make contact? Imagine listening to your favorite song. What instrument dominates the first measure? If we scanned your brain during these imaginings we would find prominent activity in visual-spatial, sensorimotor, and auditory networks, respectively, including many of the same regions that would be active during the actual experiences of navigating, kicking, and listening.

This neural simulation of experience is more than mental exhaust fumes; it's doing work. For example, research spearheaded by psychologists Roger Shepard and later Stephen Kosslyn showed that mental imagery is subject to real-world physical constraints. The time it takes you to answer a question like how many windows are visible as you walk between your front door and your bedroom depends on how long it would actually take to traverse and visually scan the space. Further, when you perform these tasks, you activate some of the same sensory and motor regions involved in actually perceiving or acting on the objects.[20] It is also well established that mental rehearsal improves performance, whether it is a speech, a surgical procedure, or a golf swing.[21] In short, thinking about experiences seems to (at least partially) reenact those experiences mentally.

Grounding embodied cognition

Score two points for embodiment. But let's consider the embodied model of conceptual representation in the broader picture. First, let's remember that it is an alternative to a particular model of conceptual representation, not an alternative to the information processing model of the mind. Processing the sensory image of that hammer and controlling the limb to use it still involves complex and multistage computational operations. The "cognitive revolution" (or information processing revolution) is still in force.

Second, even within the domain of conceptual representation it is easy to misinterpret the embodiment movement as an effort to eradicate the abstract complexities of cognition from the conceptual systems and return control to sensory and motor systems. That is, to push the theory of cognition closer to the environment, to empiricism (the view that the mind is a blank slate shaped solely by experience), and to behaviorism. This is not the case. The embodied model doesn't so much suck the meat out of the classical sandwich as it tosses out the bread. Categorization, inference, abstraction, and other "high-level" processes still occur—they have to, that's what humans do—it's just that the operations are carried out in what some scientists think of as "peripheral" input and output systems and the links between these systems.

Lawrence Barsalou, a prolific advocate of the embodied approach, puts it this way:

Three common misconceptions arise frequently about simulation/embodied/situated views. One is that they are purely empiricist with no nativist [innate] contributions. Although extreme empiricist views are possible and sometimes taken, there is no a priori reason why strong genetic constraints could not underlie a system that relies heavily on simulation, embodiment, and situatedness. . . . A second common misconception about simulation/embodied/situated approaches is that they necessarily implement [passive] recording systems and cannot implement conceptual systems for interpreting the world. . . . however, modality-specific [sensory/motor] systems can implement basic conceptual operations, such as categorization, inference, propositions, and productivity. . . . A third common misconception is that abstract concepts cannot be represented in simulation/embodied/situated approaches. Various researchers, however, have argued that mechanisms within this approach are capable of representing these concepts.[22]

Nothing has really changed, then, in terms of the complexity of the "conceptual system's" operations. It still has to infer that hammers and screwdrivers are part of the same category, will share certain features, are good for some things and not others; that a rock is different from a hammer but can be used as a hammer; and so on. But all of this "cognitive" work has been pushed down into the sensory and motor systems. In a sense, the embodied cognition folks have theorized themselves out of a job. Their proposal suggests that once we understand how sensory and motor systems work, how they categorize and infer various types of relations, we will understand how concepts are formed and represented in the mind and how sensorimotor *simulations* allow us to think and reason about those concepts. If the embodied cognition theorists are right, the field of conceptual knowledge representation now belongs to the sensory and motor scientists.

I have no problem with this idea. In fact, acknowledging that perception and action are "cognitive" domains and can perform high-level functions previously allocated by some theorists to the meat of the mental sandwich might even be on the right track. It is important to recognize, however, that calling a process "embodied" doesn't solve any problems. It doesn't tell us how the process works functionally or neurally; it only tells us in which systems it is happening and then delegates the problem solving to research in those systems. This *could* be a productive theoretical move, if the idea is correct, because situating the problem within sensory and motor systems may constrain the possible solutions. But this is a yet-to-be-tested possibility.

The same reservation holds for the favorite mechanism of the embodied movement, simulation. To say that a cognitive operation is accomplished via simulation doesn't simplify the problem,[23] it just hands it off to another domain of inquiry, in this case sensory and motor information processing. It's akin to a hypothetical rogue head-of-state who calls in his top physicists and demands that they work out how to build a nuclear weapon. The physicists come back a week later and proclaim that they've got it all figured out:

> PHYSICISTS: We have determined that Oppenheimer
> and his team have succeeded in building a nuclear
> weapon. All we need to do is simulate what they did.
>
> HEAD-OF-STATE: Great! So how did they do it?
>
> PHYSICISTS: We don't know. But simulating their
> methods will definitely work.

Saying that the brain simulates sensory and motor processes as *the* mechanism for cognitive operations involving conceptual knowledge doesn't simplify the problem or answer any fundamental questions. Rather than a revolution or paradigm shift in the science of the mind, embodied cognition might simply be characterized as the realization that sensory and motor systems perform abstract, complex computations (which sensory and motor system scientists knew already). We would be just as accurate in calling the movement the cognitivization of sensorimotor processing.

———————◆———————

SO, TO RECAP: Behaviorism held that the explanation of behavior lies mostly in the environment, with the mind contributing minimally in the form of simple learning rules. The cognitive (or information processing) revolution recognized that the environment–behavior relation is mediated in a complex way by a mind/brain that actively processes information at every level. The embodied cognition movement, often touted as a radical departure from the "classic" cognitive model, in fact is little more than the recognition that sensory and motor systems are sophisticated information processing mechanisms in their own right, capable of performing complex "cognitive" (in the colloquial sense) operations, a view quite consistent with the earliest definitions of cognitive psychology such as Neisser's.

Now we return to mirror neurons, which share a strong family resemblance to the embodied cognition movement.

EMBODIMENT, SIMULATION, AND MIRROR NEURONS

THE CLAIM that "we understand action because the motor representation of that action is activated in our brain"[24] is a theory of action understanding with simulation as the core mechanism, and which implicates a "peripheral" system as the basis of a higher-level cognitive operation. Not surprisingly, there is a great deal of mutual backscratching between the embodied cognition and mirror neuron literatures. Mirror neurons are frequently cited as a clear, neurophysiological instantiation of the central tenet of embodiment and at least one of the mirror neuron discoverers (Gallese) is now an active promoter of notions of embodiment in connection with mirror neuron function.[25]

Although there is currently plenty of love between mirror neuron and embodiment theorists, the union is not without some tension. For example, the broader embodied cognition view holds that conceptual knowledge is grounded in the sum total of one's sensory, motor, and emotional experience. One's knowledge of, say, the action concept KICK would comprise experiences with executing kicking actions, observing other people or animals kicking (which could be very different from our own kicking), and the emotions aroused by executing or observing kicking. The entire constellation of experience defines the meaning of KICK, according to the standard embodied view. Conversely, the mirror neuron variant of embodiment is highly motor centric, discounting the importance of nonmotor experience, which it considers to be semantically impoverished. If I had to pick between the two theories, I'd go with the standard embodied view because it offers a natural explanation for how we can understand actions that we can't ourselves execute.

Perhaps as a result of this tension, the theoretical reach of the mirror mechanism has seeped out of the motor system and into the broader embodied claims, muddying the overall picture. The terms *mirroring* and *simulation*, for example, are often used synonymously, particularly in social neuroscience (the study of the neural basis of social behavior) where "mirroring" emotions, intentions, pain, and touch are concepts that are bandied around with increasing frequency.[26] This is theoretically confusing because when researchers talk about functions that have little motor involvement, such as reading emotions or feeling for others, they invoke "mirror mechanisms" in nonmotor systems (we understand others' emotions or pain by "mirroring" them in our own brain) thus endorsing the broader embodiment view, but when they discuss motor-related functions they deny or minimize the importance of simulation in nonmotor systems (we understand actions by simulating those actions in our own motor system, not by simulating our sensory experiences with those actions).

In the language domain there is still more tension. The earliest claims about language and mirror neurons touted the notion that mirror neurons were important for perceiving speech *sounds* by simulating the motor programs for executing speech (see Chapter 5), and Broca's area was the simulation site. Researchers studying language from the broader embodied perspective, however, pursued the idea that the meaning of action words is coded at lower levels in the motor system. What they found in a range of experiments was that when participants read or thought about action words such as *kick* or *lick*, for example, they activate primary motor cortex corresponding to leg- and mouth-related regions, respectively.[27] Moreover, this kind of motor "simulation" affects behavior.

In one clever experiment by psychologist Arthur Glenberg and his then Ph.D. student Michael Kaschak, participants were asked to comprehend written sentences one at a time and press one of two buttons indicating whether the sentence made sense or not. The sentences of interest described a particular movement direction relative to the body such as "close the drawer" (movement away) or "open the drawer" (movement toward). Then, and here's the clever part, the researchers placed the response buttons in positions that required either a movement of the arm away from the body or toward the body. When the meaning of the sentence and the direction of the required response mismatched, subjects were slower to respond than when the sentence meaning and response direction matched. This suggested to the researchers that the comprehension of the meaning of *away*- or *toward*-related sentences depends on simulating the movements, which in turn interfered with or facilitated responses.[28] Understanding action-related meaning seems to involve motor execution circuits all the way down to primary motor cortex, according to these authors, and this form of simulation is often linked to "mirror mechanisms."

So the mirror system for language has expanded from Broca's area into primary motor cortex, which uncovers two sources of tension. One is that according to the Parma group's foundational report on the monkey experiments in 1996, mirror neurons don't live in primary motor cortex.[29] The other is that cells in primary motor cortex code more specific movements, such as movement force or direction,[30] not movement goals, such as "grasp the raisin"—the signature feature of mirror neurons that supposedly gives them the power to understand actions.

Nonetheless, given all the evidence that motor cortex activates when people think about action-related language and given that it seems like it affects behavior, we have to take it seriously. Even though we have not found any convincing evidence that the motor system is critical for speech perception or the ability to interpret the physical actions of others, maybe we will finally uncover some evidence through action-related language or, more accurately, action-related conceptual knowledge. So now let's consider whether there is evidence that action concepts are "embodied" in the motor system, as many authors propose, and, if so, whether this provides evidence for the mirror neuron theory of action understanding.

Are action concepts embodied in the motor system?

I already mentioned two of the most often cited types of evidence for motor embodiment of action concepts. One is that the motor system activates when people think about actions and the other is that thinking about actions affects our own actions. But what does it mean? We've seen how researchers in the embodied camp interpret these data, but there is another possibility.

As we've discussed already, the fact that the motor system activates while watching or thinking about actions doesn't necessarily mean that the motor system is the basis of understanding. This is true for mirror neurons in monkeys and the perception of speech sounds. It is no different with the processing of action-related language/conceptual knowledge.

Harvard psychologist Alfonso Caramazza and his former student Brad Mahon recently reviewed the literature on action-related concepts and came to the same conclusion. They pointed out that it is not surprising that the motor system becomes active when we think about action-related concepts—the concept KICK and kicking movements are related after all. It could be that when we read the word *kick* or see a picture of kicking, we understand it by accessing the nonmotorically coded abstract concept (wherever that might be coded), but that we also activate a range of related concepts and sensorimotor associations such as BALL, GOAL, BUTT, motor codes for kicking, and even motor programs for speaking the word *kick*. Now, we wouldn't say that the meaning KICK is coded in the motor programs for *speaking* the word even though they may become active (otherwise we would have to conclude that a Spanish speaker who activates *patear* would have a different concept from an English speaker who activates *kick*). So why, Mahon and Caramazza questioned, does the activation of leg-related motor regions necessarily show that the concept is coded motorically?[31]

Two separate studies published in 2013 provided further cause for doubt regarding the involvement of motor systems in conceptual knowledge of actions. One team, led by Australian cognitive neuroscientist Greig de Zubicaray, examined the fMRI activation patterns associated with listening to or reading action-related words across 15 published experiments that reported activity in motor cortex.[32] When the locations of these activations were examined in relation to a probabilistic atlas of brain regions, they found that only 55 percent of them were likely located in motor cortex. The team further compiled activation patterns from 19 published experiments in which subjects listened to or read nonsense words (such as *glemp*). Surprisingly, they found that motor cortex was more reliably activated when processing nonsense words (80 percent of activations fell within the likely boundaries of the motor system) than when processing action-related words. The implication is that it is not necessarily the *action* meanings that are causing motor cortex activity.

The second 2013 study[33] was carried about by a group at the University of Pennsylvania and also examined activation patterns across a range of published experiments, albeit with a wider net. After an extensive literature search, the researchers identified 29 reports on action concept processing involving verbal as well as nonverbal stimuli. They used statistical methods to identify patterns of activity that were reliably elicited by action concept processing across the set of studies. Although several regions were reliably activated, including portions of the posterior temporal lobe, occipital cortex, parietal cortex, and the cortex in and around Broca's area, no activity was found in primary motor cortex or adjacent tissue. While this result might be consistent with Broca's area playing some role in action concept processing, it suggests that primary motor cortex is not particularly involved.

But what about the *behavioral* effects observed by Glenberg and others? Reading a sentence that talks about pushing seems to facilitate pushing-related actions and interfere with pulling-related actions, the so-called action-sentence compatibility effect (ACE). How can the thought (concept) of PUSH interact with pushing- or pulling-related actions if the actions themselves, coded in motor cortex, are not part of the concept? Here's one possibility. Suppose there is an abstract, or at least nonmotoric concept PUSH that is wired up to a range of associations including motor programs for pushing. Now, when you think about pushing, related motor programs become activated via association, which effectively primes those actions. Given this, if your task happens to involve pushing, you are faster to execute it, not because the motor codes for pushing are part of the concept, but because the motor *associations* of the concept and the response movement happen to match.

Or think about the ACE this way. If I blew a puff of air in your eye every time I said the phrase "there is not a giraffe standing next to me," before long I could elicit an eyeblink simply by uttering the phrase alone. Furthermore, I could probably measure a there-is-not-a-giraffe-standing-next-to-me eyeblink compatibility effect by asking subjects to respond when they hear the phrase by either closing their eyes (facilitation) or opening their eyes wider (interference). This does not mean that the eyeblink embodies the meaning of the phrase. It just means that there is an association between the phrase and the action, much like the association between finger twitches and pictures of clouds, as discussed in Chapter 4. Glenberg's ACE simply hijacks an existing association between a concept and a motor program and then calls them one and the same.

Do people with motor deficits understand actions?

So we have alternative explanations for these effects. How do we decide which is correct? As we did with speech perception, we can look to neuropsychology. If the motor program for kicking is the embodiment of the meaning of the word, then damage to the motor system should disrupt comprehension of action-related words. You don't even have to do an experiment to prove that this can't be right. Again, just think about any action you can understand but you can't perform (FLY, SLITHER, or maybe SURF) or actions that you can perform and understand but don't involve your motor system (GROW, AGE, SWEAT, DIGEST, SLEEP) or abstract concepts that you can understand but seem to have nothing motoric to ground in (ARGUE, LOVE, DECIDE, BELIEVE, POST COGNITIVISM). If these concepts can be understood without grounding them in the motor system, then why is it so critical that we ground concepts like KICK, LICK, or REACH in the motor system?

A hardcore concept-grounder, such as Berkeley linguist and mirror neuron/embodiment sympathizer George Lakoff, would counter this point by saying that all of these concepts are ultimately grounded in simpler experiences that we might have had access to as kindergartners. No we can't fly but we can flap our arms while on the playground swing. Maybe you've never surfed but you have ridden on things (wagons, sleds, mom or dad's back) and you have been propelled forward by another force (being pushed on a swing or down a snowhill or just plain shoved). So you have experience with the relevant bits and this is how you understand surfing. Ditto for abstract concepts. ARGUE is built up out of experience with playground battles (*He* defended *his position vigorously but she* battled *hard and emerged from the argument* victorious) and LOVE is just a metaphor for experiences we gain on the journey home from school (*They* traveled a long road *together but ultimately had to* part ways).

What's right and wrong with the "metaphor metaphor" of meaning, as Steven Pinker called it, is beyond the scope of this book. (Pinker has a whole chapter on the topic in his book *The Stuff of Thought* that shows why the idea falls short.)[34] But I *will* comment on the *neuroscientific* viability of grounding action concepts in primary motor cortex, as some researchers would like to do. The problem is by now a familiar one. The movements *themselves*, as they are coded in motor cortex, are semantically ambiguous and therefore meaningless. The very same motor program that results in lifting a pitcher and pouring liquid into a cup could mean POUR or FILL depending on whether we conceptualize the act from the perspective of the pitcher or the cup; it could mean EMPTY if the amount of liquid in the pitcher was less than the capacity of the cup or SPILL or OVERFLOW if it was more; it could mean REFILL if the action was performed once already; it could mean COMPLY or DEFY depending on whether the action was requested or discouraged; and any of these meanings could be the result of any number of motor programs depending on whether the action was performed with the left hand, the right hand, both hands, with assistance or without, from one angle or another relative to the objects, and on and on. *The meanings simply aren't in the movements*, and the closer you get to motor codes for specific movements, such as in primary motor cortex, the farther removed from meaning you get. This is Csibra's paradox all over again.

This hasn't stopped researchers from exploring empirically the possibility of motor embodiment of action, so let's look at some of the findings. One bit of data that is frequently cited in support of low-level motor embodiment involves individuals with Lou Gehrig's disease, Amyotrophic Lateral Sclerosis (ALS). ALS is a degeneration of motor neurons that typically starts in the limbs (about 75 percent of cases) or in oral-related systems (about 25 percent of cases) and spreads from there. Language comprehension had not been documented in ALS even in patients with severe disability in the limbs or even in those with severe speech-motor control deficits. But in 2001 a group in the UK reported that a subtype of patients with this disease had difficulty comprehending words and naming pictures; verbs were particularly affected.[35]

The finding that a motor neuron disease affects verb processing did not go unnoticed by embodied cognition theorists, particularly those promoting the role of the motor system in action concepts.[36] It seems to provide just the kind of neuropsychological confirmation that the theory needs. But a closer look reveals a more complicated picture. The subtype of ALS that is associated with language problems is one in which higher-level cognitive or psychiatric symptoms are present, unlike the more common form which is primarily motor. In fact, the nonmotor symptoms are typically the first to appear in this subtype; motor involvement develops only as the disease progresses.[37]

Consistent with the presence of these higher-level deficits, the neural damage in this subtype of ALS is not restricted to low-level motor areas but also involves prominent degeneration in Broca's area.[38] But Broca's area is where mirror neurons are supposed to live. So isn't that support for motor embodiment? Not necessarily. Mirror neurons are claimed to occupy the more posterior part of Broca's area; but the more anterior part is also affected in this subtype of ALS. The involvement of *this* area is what complicates the picture because the more anterior region has been implicated in higher-level cognitive abilities.[39] This makes it impossible to infer a causal link between verb processing deficits and motor cortex dysfunction because nonmotor areas are also damaged.

A second, more recent study of *unselected* ALS patients reported a similar pattern: more trouble comprehending verbs than nouns, although even for the verbs performance hovered around 90 percent correct, and this included 14 patients thought to have a cognitive impairment beyond the motor deficits.[40] The study went further and aimed to identify the brain regions responsible for the behavioral deficits by correlating the location of cortical disease measured with *structural* MRI (diseased areas show tissue loss or atrophy) with performance on the verb and noun comprehension tests. The amount of tissue loss in primary motor cortex was not correlated with deficits on the verb tasks, nor was tissue loss in Broca's area. So dysfunction of the core of the mirror system (Broca's area) and primary motor cortex, the current focus of much embodied action concepts, does not seem to predict deficits in action knowledge. Tissue loss in premotor cortex *was* correlated with performance on the verb tests. This could be viewed as evidence for the role of this region in action concept representation. But there is a complication. Deficits on the verb comprehension tests were strongly correlated with more general cognitive decline. Given that verb-related tasks are often more difficult than noun-related tasks (actions are often *relational,* making them more complex cognitively than objects), it is possible that the "verb deficit" is really just a difficulty effect: patients with more cognitive impairment have more trouble on the harder tasks, namely, those involving verbs. The authors did not report a correlation between behaviorally measured motor dysfunction and verb comprehension.

Stepping back, the overall good performance on action-related knowledge tasks (approximately 90 percent) despite a fairly severe motor disorder like ALS suggests to me that the motor system isn't particularly critical for supporting action knowledge. To illustrate my point, consider physicist and ALS sufferer Stephen Hawking, who has severe motor dysfunction, yet understands verbs well enough to function at a very high intellectual level.

Or consider Christopher Nolan, who displayed his literary genius over a short but inspired career despite a different but no less severe motor dysfunction. U2's Bono summarizes Nolan's condition better than I could:

> We all went to the same school and just as we were leaving, a fellow called Christopher Nolan arrived. He had been deprived of oxygen for two hours when he was born, so he was Quadriplegic. But his mother believed he could understand what was going on and used to teach him at home. Eventually, they discovered a drug that allowed him to move one muscle in his neck. So they attached this unicorn device to his forehead and he learned to type. And out of him came all these poems that he'd been storing up in his head.[41]

Nolan, who had cerebral palsy, described his own experience in 1987:

> My mind is like a spin-dryer at full speed, my thoughts fly around my skull while millions of beautiful words cascade down in my lap. Images gunfire across my consciousness and

while trying to discipline them I jump in awe at the soul-filled bounty of my mind's expanse.[42]

The literary world was surprised with the eloquence that flowed out of such a crumpled body. A *Publishers Weekly* review of Nolan's first novel, *The Banyan Tree,* states:

Occasionally a book comes along that is so innocent and seemingly unmindful of current literary fashion that, paradoxically, it shocks. . . . So will this extraordinary first novel by Irishman Nolan, who, at 33 years of age, has spent his entire life as a quadriplegic, unable to speak. You wouldn't know it from his perfectly crafted, exquisitely written story.[43]

He has spent his life as a quadriplegic *but you wouldn't know it,* as if the inability to speak or indeed control his body should necessarily limit his ability to understand the world and, given a technological channel, to communicate his understanding exquisitely. Interestingly, Nolan told *Publishers Weekly* that the idea for the novel was sparked by the image of "an old woman holding up her skirts as she made ready to jump a rut in a field."[44] In the context of theories that promote the motor system as a critical part of the ability to understand actions, it is telling, I think, that a quadriplegic author could take inspiration for a critically acclaimed novel from an image involving two actions that he never performed himself.

There are many examples of individuals with cerebral palsy who cannot speak or control their bodies yet succeed in understanding and contributing to the world. Some, like Nolan, have become famous for their work; many, many others have had less public successes.

Embodied theorists point to other evidence in support of the idea that the motor system plays a critical role in action understanding. Individuals with left frontal lobe lesions, for example, often have more trouble naming actions than they do naming objects.[45] However, the source of the deficit does not seem to involve the concept for the action. More often than not, patients with frontal lesions and deficits in naming actions can comprehend the same actions (e.g., RUN or PUSH) quite well, or they exhibit performance differences depending on whether they are speaking or writing the name, suggesting a deficit that is modality- or task-specific rather than concept-general.[46] When naming and comprehension of actions do co-occur in stroke patients, the brain regions that are affected tend to be in the superior temporal lobe rather than in motor-related areas, which implies a nonmotor explanation for the source of the deficit.[47]

TMS provides another source of data. A recent study found that stimulation of the hand area of left motor cortex (controlling the dominant hand in the right-handed participants) affected how long it took for subjects to recognize written words denoting dominant-hand-related actions such as THROW, WRITE, ERASE, COMB, SLICE; stimulation of the hand area of the *right* motor cortex, which controls the non-dominant hand, did not. The authors concluded that "processing an action verb depends in part on activity in a motor region that contributes to planning and executing the action named by the verb."[48] In other words, the ability to read the words *write* and *throw* is dependent on the motor cortex that directly controls the hand that writes and throws. There are three reasons why this conclusion doesn't follow from the results of the experiment.

First, the study did not control for response bias, a complication we encountered in Chapter 5. Subjects were aware that the experimenters were targeting their motor system and they were aware of whether their left or right motor areas were being stimulated. Further, it wouldn't take much to notice that half of the words they were reading related to actions performed by the hand. Thus, independent of the TMS stimulation itself, the experimental conditions could have tipped subjects off as to the purpose of the experiment, which could have influenced their response. In this respect it is interesting that motor stimulation to the left-hemisphere region *facilitated* response times, when the type of stimulation they used usually interferes with the function of the stimulated site. The authors explained this paradoxical result as having to do with opposite-going effects at the neural and behavioral level: interrupting neural function could have facilitated behavior. However, another possible explanation is that the behavioral effect—faster word recognition with dominant motor cortex stimulation—has little to do with interrupting conceptual representations.

Second, even assuming that the facilitation effect was caused by TMS motor stimulation, it is impossible to know whether the stimulated region itself codes a part of the conceptual knowledge of the action or whether it is simply connected to the network that codes the conceptual knowledge of the action, like the association built between pinky movements and clouds. Suppose that the concept JUMP is coded somewhere outside of the motor system, which is the "classical" assumption. As noted earlier, it is surely the case that an association exists between the concept JUMP and the motor program(s) for jumping. For example, imperative statements—"Jump!" "Duck!" "Shoot!"—are effectively asking the listener to execute the motor programs that correspond to the meaning of the word. Given such associations, it is possible that stimulating the motor system with TMS simply activated nonmotor associated concepts, which facilitated word recognition in the experiment.

Third, even if the effect is caused by the stimulation *and* the motor code is part of the concept, we still have to consider how much of a role that motor code is actually playing. Consider the following three verbs that were used in the experiment: *to erase, to comb, to slice.* How much of the meaning of those verbs depends on the involvement of dominant-hand movements? Erasing is erasing no matter whether you do it right-handed, left-handed, with your feet or your backside. Or think of it this way: the experiment utilized 96 verbs that referred to actions performed primarily by the dominant hand. If the dominant-hand motor involvement is an important part of the meaning of these verbs, they should form a natural semantic grouping because they have something important in common. Words like *apple, banana,* and *watermelon* have important and meaningful commonalities and so form a natural semantic class, FRUITS, as do *chinchilla, alligator,* and *giraffe,* ANIMALS. But we don't naturally group dominant-hand actions as a semantic category like we do fruits, animals, tools, or musical instruments. If you think of apples, you very well may automatically think of bananas or the category FRUIT because they share semantic features. But if you think of erasing do you automatically think of slicing (a dominant-hand action)? Not so much. If "dominant-hand" plays a role in word or concept processing, it is a miniscule role indeed.

If we want to understand what comprises the meanings of actions we need to look well beyond the motor system. *The meaning simply is not in the movement.*

The motor system, though, may have some role to play in coding the meaning of actions. I for one learn (and therefore presumably understand) much better by *doing* than watching or following, whether it is learning how to get somewhere, learning a dance step, or crunching through a data analysis stream. Controlled experimental evidence confirms this introspection, which I suspect rings true for many of us.[49] The question is *why* does doing improve learning and understanding, and does a facilitation of learning mean that the code underlying that knowledge lives in the motor system? This is a topic that we take up in the next chapter.

What does "mirroring" mean in the context of information processing?

As I pointed out previously, to claim that a process is accomplished by simulation doesn't solve any problems. It is worth considering mirror neuron claims in this light. Here is a recap of the hypothesis from a recent review article by Rizzolatti and philosopher Corrado Sinigaglia (the emphases are mine):

> From the discovery of mirror neurons, the interpretation of this finding was that they allow the observer to understand *directly* the goal of the actions of others: observing actions performed by another individual elicits a motor activation in the brain of the observer similar to that which occurs when the observer plans their own actions, and the similarity between these two activations allows the observer to understand the actions of others *without needing inferential processing.*[50]

This is a classic embodiment-type statement: a complex cognitive operation, understanding, can be carried out "directly" without all those complications like "inferential processing" that are classically ascribed to the cognitive system. But this doesn't get us very far:

> MIRROR NEURON THEORISTS: We have determined
> that when we carry out an action of our own we
> understand the meaning of that action. When we
> observe someone else's action, all we need to do is sim-
> ulate that action in our own motor system and we will
> achieve understanding.

> SKEPTIC: Yes, but how do we understand the meaning
> of our own actions in the first place?

> MIRROR NEURON THEORISTS: We don't know, but we
> know that simulating that process will work.

Computational neuroscientist Michael Arbib has made similar points. Arbib, you recall, was coauthor with Rizzolatti on one of the early theoretical extensions of mirror neuron theory into human cognition. Their 1998 paper, "Language within our Grasp," made the case for the role of mirror neurons in human language (discussed in Chapter 3) and kicked off a decade of work on the topic. Being a computational neuroscientist, Arbib's work focuses on developing functioning computer models of neural systems and so he has to worry about what it means *computationally* to say that a system uses simulation, direct matching, resonance, or mirroring. *Conceptual models*, verbally laid out theories like those we've dealt with in this book, don't have to be as explicit. In a recent review,[51] Arbib and colleagues write:

> A general pitfall in conceptual modeling is that an innocent
> looking phrase thrown in the description may render the model
> implausible or trivial from a computational perspective, hid-

ing the real difficulty of the problem. For example, terms like "direct matching" and "resonance" are used as if they were atomic processes that allow one to build hypotheses about higher cognitive functions of mirror neurons.

Arbib still believes mirror neurons are contributing to action understanding, but his computational work had led him to a more moderate perspective, which in my assessment is more plausible and interesting. We return to this in Chapter 10.

The exact same pitfall haunts all the nonmotor forms of "mirroring" as well. We feel someone's pain by simulating that pain in our own sensory system. We read people's minds by simulating their situation within ourselves. The result of this kind of theorizing is that the whole brain has become one big simulation, which means that *simulation* as a mechanism has lost its explanatory power. It is now simply being used as a synonym for terms like *information processing* or *computation*. Same old problems, new vocabulary. As philosopher Patricia Churchland remarked when I pointed this out in a lecture at UCSD (and I paraphrase), "Hey I recognize that problem!" She said, "That's the how–does–the–brain–work problem!"

We've seen this trick before. To paraphrase Chomsky's critique of Skinner:

> This [theoretical] device is as simple as it is empty. Since *simulations* are free for the asking . . . we can account for a wide class of *complex cognitive operations* in terms of *embodiment theories* by appealing to *simulation*. But the word *simulation* has lost all objectivity in this usage. *Simulations* are no longer *simple operations*; they are driven back into the *complexities of high-level cognition*. . . . It is clear from such examples, which abound, that the talk of *simulation* simply disguises a complete retreat to *abstract, computational cognitive* psychology.

This analysis is perhaps somewhat overly harsh; knowing that we should be looking for abstract conceptual representations in sensory and motor systems may help develop new and better models of how concepts are stored and put to use in the brain. This is progress and I'm fully onboard with the idea of exploring it. But it is incorrect to imply that punting on the question of how we understand actions by appeal to a mirroring or simulation mechanism solves the problem. It doesn't. It just hands the ball off to another team.

Although we don't yet know exactly what that other team looks like, the next chapter introduces some of the players.

7

Feeling, Doing, Knowing

THE FEELING OF KNOWING

I MAGINE FALLING asleep one night and waking up in a nightmare. As you open one eye of consciousness you notice that your right arm is "asleep." Annoying, but nothing to be alarmed about. All you need to do is roll over, wait a few seconds, and the tingle of life will rush back into your arm. But when you try to roll over, you find that you can't; in fact, you can't move at all except for your head. And that's when you notice that your whole body is "asleep." From the neck down you can't feel a thing, not even the bed beneath you. You feel as if you are floating. Is your body even there? You look down and find that it is. You wait a few more seconds hoping the feeling and the ability to move returns. The seconds turn to minutes, then hours, then a lifetime.

For Ian Waterman, a then nineteen-year-old butcher living on the UK island of Jersey in the English Channel, this was no nightmare. Ian was in the hospital fighting a severe virus when one morning he woke up and couldn't feel or move his body. Doctors had no idea what was going on, what his prognosis was, or how to treat him. He recovered from the virus but his body remains lost to him. In a BBC documentary called "The Man Who Lost His Body," Ian described his tragic experience poetically:

> Turned every 2 hours like a joint of meat. Basted with lotions. Unmoving like a statue. Mind filled with emotion. Limbs dead to the touch, movement impossible. Lying on a bed, eyes fixed on a flaking ceiling, wishing those flakes would turn to cracks and the ceiling fall to take me from this misery. What use an active brain without mobility?

But Ian didn't lose mobility completely. In fact, his motor system is virtually unaffected. What is lost is his ability to feel, a condition known as *large-fiber sensory neuropathy.* Ian's case is extraordinarily severe. He lost all sensation from touch as well as proprioception, sensory signals from receptors deep within the muscles and joints that signal body position. Nerve fibers carrying pain signals as well as motor signals are largely unaffected. Cases like Ian's tell us that the ability to *move* depends on the ability to *feel.*

Most people with Ian's condition are unable to walk or care for themselves without assistance. Ian's doctors predicted the same outcome for him. But Ian disagreed and with heroic effort taught himself to walk, drive, work, and live independently again. He did not somehow retrain his brain to feel his body again; that ability is gone for good. What he did, and continues to do every waking minute, of every day, was to use *visual* feedback in place of the normal and automatic *somatosensory* feedback. When you or I walk, we can look around, peel a banana, think about a recent conversation. The walking itself is effortless and automatic; we don't have to think about shifting weight to one leg, lifting the other, pulling it forward, and so on. Nor do we have to look down to know what our legs are doing. This is because we can *feel* our legs and this information is what the brain and spinal cord use to coordinate walking. Ian lost this proprioceptive information, however, and so can only will his legs to move if he is looking at them and consciously planning each movement. Dual tasking while walking is impossible. The same is true for his hands. When he can't see them, they tend to "wander," sometimes knocking things down or smacking into someone.

Ian's doctor describes his movement control abilities in more detail:

> All [Ian]'s useful movements require constant visual control and mental concentration. In the dark he is unable to move and such is his requirement to concentrate on movement that he cannot daydream whilst walking. He finds now that walking on level terrain requires about half the amount of concentration that it did. Walking on uneven ground still demands full concentration. . . .
>
> His mental concentration is finite. If in a chair he can pick up an egg without cracking the shell. But he would not be able to walk and hold the egg; for then his concentration would be on walking and his grip on the egg [would] become too firm to avoid breaking it. He drives a car with hand controls and finds it very relaxing. Here he has his hands frozen around the controls which remain in his peripheral vision, though his concentration is on the road. . . . All movement tasks appear to require continuous visual feedback, except for freezing in postures which are then held with frequent visual checks.[1]

Ian is completely dependent on visual feedback. Turn off the lights and he collapses to the ground, according to the BBC documentary.

There are milder forms of sensory nerve degeneration, such as *tabes dorsalis*, a condition caused by untreated syphilis. Affected patients develop a stomping gait because they lose sensory information telling them when their feet contact the ground. By stomping they can generate vibrations that reverberate all the way up to the trunk, where sensation is better preserved and ground contact can be detected. You've also likely experienced a form of sensory loss after a lidocaine injection at the dentist and may have noticed its effect on your ability to speak.

The fact that somatosensory feedback is a critical part of motor control is not new news. British neurologist and physiologist Henry Charlton Bastian wrote on the topic back in 1887, stating, "It may be regarded as a physiological axiom, that all purposive movements of animals are guided by sensations or by afferent impressions of some kind."[2] Experimental work over the decades found that blocking somatosensory feedback from a monkey's limb (while leaving motor fibers intact) causes the limb to go dead. With training the monkey can learn to reuse it clumsily, but only with visual feedback; blindfold the animal and motor control degrades dramatically.[3]

This illustrates a critical point: without sensory information, the motor system is literally and figuratively blind. We don't naturally think of movement as being so dependent on sensation, but give it a moment's thought and it becomes obvious. How do you know where to reach for an object? Because you *see* its location. How do you know in which direction to reach? Because you *feel* where your hand is in relation to the object. How do you know whether to grasp it with the whole hand or with a pincher grip? Because you see the size and shape of the object. How do you know whether you have succeeded in grasping it? Because you feel it and see it in your hand.

Imagine you live in Ian's somatosensation-free world, but to an extreme degree such that you can't feel a thing, including pain or even your head position. Your only sense is sight. You are sitting at a table with your arms resting on the surface in front of you; a tennis ball sits just in front of your hands. Now close your eyes. You cannot feel the table supporting your arms. You cannot feel the chair under you. You cannot feel your feet on the floor. You feel like you are floating in space. Now raise your hand above your head, hold it for five seconds, and replace it on the table. You command your arm to move but you cannot feel it move. How do you know when to stop the upward movement? You don't. How do you know what forces to apply to keep it balanced up there? You don't. If you don't know where it is exactly, how do you apply the forces to return it to the table? Without sensory feedback of some kind, you have no idea. Now, with that same arm—wherever it is—reach for the tennis ball. How do you know which way to reach? If you happened to reach in the right direction, how do you know when to clasp your hand? How do you know if your hand is even open or closed? How do you know you're touching the ball? Have your attempts to move generated any movements at all? You don't know.

If we could play out such a scenario, most likely when you first close your eyes your body would start listing in the chair; you wouldn't be able to feel the sway and so you wouldn't be able to correct it and you would fall to the floor. You wouldn't know you fell, though, because you would not be able to sense the fall or the impact and all of your attempts at movement would be jerky, hapless fits, if that. Sensory information is, literally and figuratively, the eyes and ears of the motor system, allowing it to "know" what it is doing. To state it in the vernacular of the embodied cognition proponents, the motor system is "grounded in sensation."

Sensory grounding in speech

The same sensory-dependent situation holds for motor speech behaviors. Individuals born deaf experience great difficulty learning to speak, even though they have no difficulty learning language in the form of manual signs. Why is this so, given that they have normal muscular control over their vocal tract? Because they can't hear the targets of speech acts, namely, the *sounds* of words. Imagine trying to learn to speak a new language that you can't hear! (Some deaf people can speak quite intelligibly; the actor Marlee Matlin comes to mind. But note that this is probably related to the amount of residual hearing the individual has[4] and how well he or she can rely on somatosensory information—the ability to feel the position of the tongue, lips, and larynx—which is available to deaf people.)

The role of auditory information in speech is also evident in hearing/speaking individuals. If you've ever heard your own voice echoing back with a slight delay on a bad phone connection, you know how disruptive the wrong kind of feedback can be on your ability to speak. Such *delayed auditory feedback* of one's own speech is well known to decrease speech fluency. Or you may have noticed that a word you hear spoken on the television or radio in the background of a conversation suddenly finds its way, quite unintentionally and nonsensically, into the middle of your own sentence. Or, less fleetingly, you may have noticed yourself or an acquaintance picking up an accent after time spent with talkers from a different region of the country, a phenomenon known as *gestural drift*, the largely unconscious tendency to shift speech patterns toward those that you hear around you.[5]

A rather dramatic example of gestural drift for me is an intonational pattern in North American English called *uptalk*, the tendency to end nearly every sentence with a rising pitch, as if it were a question. Canadian psychologist Hank Davis called uptalk a "nasty habit," "the very opposite of confidence or assertiveness," an "epidemic," and the "meme from Hell."[6] Science writer James Gorman caught the uptalk bug early and gave it its name in a 1993 *New York Times Magazine* piece:

> I used to speak in a regular voice. I was able to assert, demand, question. Then I started teaching. At a university? And my students had this rising intonation thing? It was particularly noticeable on telephone messages. "Hello? Professor Gorman? This is Albert? From feature writing?"
>
> I had no idea that a change in the "intonation contour" of a sentence, as linguists put it, could be as contagious as the common cold. But before long I noticed a Jekyll-and-Hyde transformation in my own speech. I first heard it when I myself was leaving a message. "This is Jim Gorman? I'm doing an article on Klingon? The language? From 'Star Trek'?" I realized then that I was unwittingly, unwillingly speaking uptalk.[7]

In all these cases information impinging on your ears is messing with the auditory targets of your speech gestures, like someone fiddling with the position of a dartboard in the middle of a game. In the first case, delayed auditory feedback disrupts the timing delay between when your brain sends the speech movement commands and when the auditory feedback is expected, therefore tricking the system into thinking that a motor programming error has occurred and needs to be fixed. In the second case, the background TV dialogue puts an external target in front of you and you reach for it reflexively, so to speak. In gestural drift, being surrounded by examples of a particular speech pattern (or auditory target) eventually replaces your old targets with new ones, thus changing your own speech pattern. (Targets can be stored sensory memories as well as physical objects in the environment, and like most memories they are subject to modification.)

Parenthetically, despite claims to the contrary, uptalk is not a reflection of self-doubt or indecisiveness, as pointed out by linguist Mark Liberman.[8] The popular interpretation assumes that uptalk statements are questions, but they're not. They are something closer to a contraction. Uptalk statements might have started out as something like, *My students had this rising intonation thing. You know?* but then *You know?* got dropped and the question intonation shifted onto the end of the statement. To understand uptalk's function, just think about what function *You know?* serves. It is to control your listener's attention and to make sure they are with you in the conversation. It has more to do with control than indecisiveness.

No matter where we look, action is firmly and indispensably grounded in sensory systems. Is the reverse true? Are sensory systems indispensably grounded in action? Not so much. We've seen example after example of how we can understand the world without the ability to act. You don't need to be able to speak to understand speech. You don't need to be able to smile to understand smiles. You don't need to be able to fly to understand flying. While the motor system is absolutely dependent on sensory systems to get anything done, the reverse is absolutely not the case. The relation is asymmetric.

KNOWING THROUGH ACTION

THIS IS not to say that the ability to act in the world doesn't make *some* contribution to perceiving and therefore understanding. On the contrary, action contributes a great deal. At the most basic level, without the ability to move around in our environments, the scope of what we can experience perceptually would be extremely limited. Normal vision, for example, depends on the ability to orient toward interesting features in visual space. Without the ability to move our eyes, head, and body, we would be perceptual slaves to whatever happens to fall directly in the middle of our gaze. Basic hearing is less movement dependent but still benefits from the ability to shift the position of our ears in space. And most of the touch sensing we do is a direct result of movements that put our body into contact with objects. In short, perception is highly active.

This last point was championed by the late psychologist James Gibson, who argued that we need to consider perception in its natural habitat rather than in contrived laboratory conditions.[9] For Gibson, a visual scientist, this meant studying vision in the context of action. In fact, he argued that in the real world the line between perception and action is blurred, even indistinguishable. We don't so much see objects, we see opportunities—*affordances*, he called them—for action. We don't see chairs; we see opportunities for sitting. We don't see cups; we see opportunities for grasping or drinking. The whole point of perception, according to Gibson, is to afford action. Of course, we could turn this around and say the whole point of action is to generate more desirable sensations: relieving fatigue, quenching thirst, and so on.

Gibson's work is popular these days with embodied cognition and mirror neuron theorists because he pushed to break down traditional barriers between domains of cognition and because he emphasized the role of action in perception. But we should not confuse the fact that action is an important part of perception with the idea that motor representations alone are the basis of perceptual understanding. Walking the rim of the Grand Canyon certainly broadens one's perception and understanding of the place, but this does not mean that you neurally code your Grand Canyon experience in the form of leg movements, head turns, and eye scans. Action critically broadens the scope of perception, but it doesn't define it.

Another common argument for the critical role of action in perceiving is the simple fact, alluded to in the last chapter, that we tend to learn best by doing, and having done it ourselves, we tend to understand it better. Neuroscientist, tennis player, and mirror neuron enthusiast Marco Iacoboni argues this point with respect to watching a tennis match. He claims to understand the game better than his non–tennis-playing mom.[10] Many of us no doubt have the same intuition for sports or other activities that we have ourselves participated in. Why? It is natural to look to the movements themselves as the answer. But our own movements are triggered by sensory targets and result in sensory feedback. So having more experience *doing* necessarily means having more experience *sensing*. Someone who plays basketball has immediate and personal experience with various formations and movement trajectories, all of which provide sensory targets for passes, shots, blocks, picks, and lanes to drive to the basket. Executing those actions results in sensations of changes in body position, the feel of the ball in your hands, pressure against an opponent during a block or a pick, and so on. Having experience in performing an action *can* lead to a better understanding of those actions, but this does not imply that the meaning is stored in the motor system or that motor simulation is necessary to achieve understanding.

I explained this point to a science writer once who remained unconvinced. She countered that she plays a lot of basketball and when she watches a game she finds herself responding to what she sees physically, leaning this way or that, making miniature passing or shot movements, and so on. "Isn't that a reflection of my motor system mirroring the movements I'm seeing," she asked, "thus augmenting my understanding?" I don't think so. Your motor system is resonating with the array of *sensory* targets presented to you. The reason why you tend to respond motorically, whereas someone else who doesn't play basketball might not (even if she has a similar understanding of the game), is that you have a great deal of experience with similar sensory situations and have strong associations between particular situations and effective motor responses. Like a conditioned eyeblink, you can't help but trigger appropriate motor responses.

Consider the common male reaction to watching another man get kicked in the groin, namely doubling over and covering up. Notice that he doesn't simulate the kicking action. Instead, the observer "simulates" the response of the kickee because he's observed similar kicking actions before and knows what it feels like when the kick connects. His doubling over is a reflexive response appropriate to avoiding pain in such a situation. Motor responses triggered by observing the action in this scenario, as in sporting events, are not the *cause* of the richer understanding while watching, they are a *consequence* of understanding the scenario more richly, which often boils down to having more sensory experience. Now, there *are* situations in which people tend to mimic the actions of others, often unconsciously, a topic we consider in the next chapter.

So one reason why *doing* can augment observational understanding is that more action experience equates to more perceptual experience. There's another potential mechanism though. Actually performing a task can make you think of or notice things that you wouldn't otherwise. Take surfing, for example. Someone who hasn't surfed understands that you paddle out toward the waves, turn the board around, and then paddle *with* a wave. When it propels you forward, you stand up, turn, and hang a few off the nose. First-time surfers almost uniformly position themselves too far back on the board when they are paddling, such that the nose of the board is angled toward the sky. This keeps you from catching waves because the board is *pushing* water rather than skimming on top of it. An experienced surfer knows to pay attention to the position of the nose relative to the water line while paddling. This same surfer can sit on the beach, watch two different paddlers trying to catch a wave, and tell you which one is likely to catch it based on board angle. Action experience can tune the senses by highlighting what's important to attend to and what isn't. This is not a motor thing per se, it is a sensory/attention thing that is shaped by acting in a particular environment.

This is not to say that the motor system doesn't store any useful information. There is such a thing as motor memory or motor knowledge. There is no question that motor skills can be honed to an exquisite level, where a task (e.g., driving, typing) can be executed "on autopilot" or "in your sleep," that is, with little sensory feedback. These skills are, indeed, stored in motor systems and constitute a kind of knowledge, or "motor understanding" of the action. One might, therefore, be tempted to argue that action understanding is incomplete without such motor knowledge. For instance, you may not know how to play "Blackbird" on a guitar, so in that sense there are things that you don't "understand" about the song. But, again, the song isn't defined by how you move your fingers on a guitar; it's something closer to an acoustic track of the melody. One could play it on a piano and it would still be "Blackbird." Or one could play it on an untuned guitar—same movements, no longer "Blackbird." Important information is stored in the motor system, but that information is motoric, not meaningful (semantic) information, as we see every time we take a minute to scratch the surface of a motor-based claim of meaning.

The point of this discussion is that while indeed a tight coupling exists between perception and action as the mirror neuron theorists claim, they are placing the interpretive load, the semantics, in the wrong place.

The natural history of motor theories of perception reflect this error. They start out as implicating low-level motor programs and then get drawn back toward higher levels of abstraction as the gravity of the evidence mounts. The motor theory of speech perception began as a claim that motor codes for controlling speech movements were responsible for our perception of speech sounds. Confronted with clear evidence that this can't be the case, the theory supporters revised it a bit and argued that it wasn't the low-level motor programs but more abstract "*intended* gestures" that did the perceptual work.[11] But then you have to ask, what is an intended gesture? Is it the intention to move the mouth in a particular way or the intention to make a particular *sound*? Research on motor control in speech production tells us that it is the latter, that the goals of speech acts are acoustic.[12] So by abstracting away from motor particulars, one pretty quickly finds oneself in sensory systems.

Mirror neuron claims have followed a similar trajectory. At first, simulation of the movements themselves via congruent mirror neurons enabled understanding. Now the emphasis is on neurons coding abstract motor "goals" quite independently of how the goals are achieved motorically. Again we have to ask, what are those goals? Are they particular movements (extend arm, open hand) or are they sensory states (hand in same location as object with hand grip shape matching object shape)? The discussion regarding the critical role of sensory systems in guiding action points to the latter.

Viewing the world through mirror neurons

We can see that recent mirror neuron accounts of action understanding are just sensory theories in disguise if we look closely at what mirror neuron supporters claim. Take Iacoboni's description of what his enhanced understanding of Roger Federer's tennis swing actually entails:

> When I watch the same backhand volley, my internal motor knowledge of how to hit a backhand volley gives me a much richer understanding . . . of that action. That is, I can typically predict ball direction, speed, and placement—even a few moments before Federer's racquet hits the ball—according to racquet orientation, body position, and motion.[13]

His enhanced understanding emerges not in terms of movements but in terms of sensory states (ball trajectory and speed) predicted on the basis of other sensory states (racquet orientation, body position). If what you are trying to predict is sensory, and if the relevant information for generating that prediction is sensory, why bother simulating anything with the motor system? Why not just learn associations directly between sensory states, such as the relation between observed racquet orientation and resulting ball trajectory?

There is a good reason why Iacoboni describes his enhanced knowledge in sensory terms: the relevant generalizations are not motor. Consider the most studied action of the mirror neuron crowd, GRASPING. As pointed out correctly by mirror neuron theorists themselves, one can grasp with the hand, the lips, the arms, the feet, any number of tools or machines, or the mind. And one can grasp large or small objects that are high or low, to the left or right, and on and on. Motor theorists like to say that our understanding of all these kinds of grasping is "grounded" in the motor system. But what motor representation could it possibly be grounded in to cover everything from a bird grasping a worm to a thinker grasping an idea? Embodied cognition theorists like to say that it is all of them and the sum total of all of this grasping defines the concept (or, alternatively, that grasping with the hand is a kind of concrete conceptual anchor and all other instances of grasping are understood as generalizations of this basic form). Fine, so how does the brain know what to include in the summing operation (or what to generalize the basic form to)? It can't be based on similarity of movement because very different movements (or no movement) can result in grasping. To find what is common among all the types of grasping, you have to leave the motor system again because nothing there naturally groups them, as we saw in the last chapter. What binds together all the examples of grasping is more abstract than any particular movement. We are back to square one. Deferring to the motor system doesn't solve any problems.

The fact that I'm even writing this book suggests that not everyone agrees with my arguments that minimize the role of the motor system in understanding the world. In a recent theoretical paper Arthur Glenberg and Vittorio Gallese explain the motor-centric alternative to my position:

> We believe that the basic function of cognition is control of action. From an evolutionary perspective, it is hard to imagine any other story. That is, systems evolve because they contribute to the ability to survive and reproduce, and those activities demand action. As Rudolfo Llinas puts it, "The nervous system is only necessary for multicellular creatures that can orchestrate and express active movement . . . ." Thus, although brains have impressive capacities for perception, emotion, and more, those capacities are in the service of action.[14]

I disagree. The basic function of cognition is not the control of action. It is to increase the probability of surviving and reproducing (that part they got right). Systems involved in "perception, emotion, and more" are not in the service of action. They are in the service of survival and reproduction. Glenberg and Gallese are wide-eyed to the fact that sensation isn't very useful without action, quite correctly so, but they ignore the fact that action is useless, indeed impossible, without sensation. Cognition, in the broad sense including perception and action, is a system, the whole of which evolved under selectional pressures. Evolution did not, could not, single out the motor system alone to shape.

WHERE DO WE UNDERSTAND?

GIVEN THAT the motor system fails to capture the semantics of actions, one might be tempted to infer that meaning arises in sensory systems. This is not the case either, at least not in any simple sense. All of the problems we encountered on the motor side also hold on the sensory side. When we see a bird grasping a worm or a shopper grasping a can of soup we might ask, what perceptual features define the action such that we recognize both as instances of grasping? Perceptually the two events are quite different.

This is a general problem in perceptual science. Think of all the physical variability in size, shape, color, and location of the objects we categorize as cups or chairs or dogs; then think about all the different angles you can view these objects from; then think about how many ways parts of these objects can be obscured or occluded by other objects or by shading, yet we still recognize them. Further, we can recognize objects by sight or by feel or by sound alone. Categories of actions (GRASPING, JUMPING, KICKING) or objects (DOGS, CHAIRS, VEHICLES) can't be defined by low-level sensory features any more than they can be defined by low-level motor features; there's too much variability at that level of analysis. The core of the meaning of an action or an object has to be something more abstract that cuts across low-level physical details and even sensory and motor systems.

And then there is the lot of abstract concepts such as ABSTRACT, CONCEPT, SUGGEST, to name three from the previous paragraph, which are difficult to tie down to sensorimotor systems. This is one reason why classical theories of conceptual knowledge promoted an abstract system that is not tied to a particular modality of sensing (seeing, hearing, feeling) or acting.

A framework for conceptual knowledge

So where do we stand? It's a confusing state of affairs. I've argued throughout this book that motor systems fail to capture how we understand the world, including how we understand actions. I tried to illustrate in the first part of this chapter that sensory systems are absolutely critical for giving the motor system purpose: they provide targets for actions and critical feedback on the consequences of actions. This hints that we should look to sensory systems if we want to find the neural basis of understanding. But there is also good reason to believe that conceptual knowledge systems abstract across sensory details and indeed sensory modalities. We need to develop a theory that transcends sensory and motor systems, which leads us to a classical sandwich model of conceptual organization. But, as we saw in the last chapter, this is probably too strong a position; some sensorimotor details clearly impact how we understand and reason about the world, so we need a theory that is at least partly "grounded."

I'm not going to pretend that I have the answer for how the brain represents and processes conceptual knowledge for actions or any another other category of information—philosophers and scientists have been working on it for centuries with no resolution—but I believe that by situating these problems in a broader framework of brain organization we can at least see the outlines of a solution.

That broader framework includes two principles of organization that have been persistent themes in the history of neuroscience. One is that the nervous system is hierarchically organized. The other concerns the brain's two different uses of sensory information as we discussed in Chapter 4: for the immediate control of action (*how*) and for understanding (*what*). We consider these in turn.

PRINCIPLE I: HIERARCHICAL ORGANIZATION

Reflexes are controlled by relatively simple sensorimotor circuits handled at the level of the spinal cord. The knee–jerk reflex, for example, operates like this: the tap of a doctor's hammer stretches the tendon in the knee which stimulates sensory receptors in the leg muscle→these sensory cells synapse with motor neurons in the spinal cord both directly and indirectly (via interneurons, also in the spinal cord)→spinal cord motor cells send signals back to the muscles, causing the leg to move. Reflexes are very useful for preventing injury due to sudden and unexpected pokes, stretches, and burns, but they are really bad at preventing you from dropping a hot plate on the floor when you start to feel the heat in your hand. To override reflexes, and therefore allow for more *flexible* behavior, we need higher-level circuits that can understand that dropping the plate ultimately causes more grief than holding on to it. Several levels of sensorimotor circuits have been identified, making loops and side loops through the brainstem all the way up to the cerebral cortex.

If we look within sensory or motor systems in the cortex we find even more hierarchical layering.[15] Cells in early cortical stages of visual processing, such as in primary visual cortex, respond nicely to simple stimulation such as bars of light in particular orientations presented in particular places within the visual field; they are tuned to relatively low-level visual features. Move several processing stages further into the cortical system (into other visual areas) and we find that cells respond preferentially to more complex visual stimuli such as faces or other object-like stimuli. The precise position of the object in the visual field matters less, as does the size (retinal image size, that is) of the stimulus, which strongly suggests the existence of a higher-level code or representation that is abstracted away from details—a face is a face whether it is to your left or right, far or near. The connectivity pattern between various regions in these systems also supports the notion of a hierarchical organization.[16]

PRINCIPLE 2: DUAL PROCESSING STREAMS FOR *what* VERSUS *how*

We discussed the idea that the brain carries out two different computational tasks with the same sensory input—recognizing *what* is being perceived and recognizing *how* to react motorically to it— in Chapter 4 and mentioned in other places in this book. The gist is that the sensory properties that matter most for *what* versus *how* are different, indeed complementary, which is why we need two systems, one for each computational task. To recognize a cup as a cup the visual system needs to ignore absolute size, orientation, details of its shape, distance from the observer, and so on. Rather, recognition requires the system to extract the *invariant* features that make a cup a cup. To reach out and grasp the same cup, however, the visual system needs to pay close attention to the particular details; extracting only invariant features does not help in coding the trajectory of the arm toward the location of the cup or in orienting and sizing the hand appropriately. Further, a successful reach and grasp doesn't require that the object be recognized. If Sir Isaac Newton could slip through a wormhole and land in my living room he could just as successfully reach for my remote control as I could even though he would have no clue what it is. And again, by way of refresher, the *what* and *how* streams are largely segregated in the brain into a ventral, temporal lobe stream and a dorsal, parietal lobe stream, respectively.

How do these two principles frame a possible solution for the problem concerning the nature of conceptual knowledge? Let's start by clearly stating the problem, which stems from a tension between the need for abstraction in conceptual knowledge and the many observations that at least part of our knowledge is not distinct from sensory and motor experience. This tension can be summarized in a few ways. Here's one:

- On one hand, a theory of conceptual knowledge of actions or any other type of information that is completely divorced from sensory and motor systems is too strong a position. When we recall what we know about lions or Led Zeppelin songs or how to surf we can't help but relive at least some of the associated sensory and motor experiences, as clever behavioral and brain imaging studies have shown.
- On the other hand, what really defines these concepts isn't tied down to particular experiences; the core meaning is abstracted across sensory and motor experiences. We can recognize stylized headdresses in the theatrical version of *The Lion King* as examples of lions or even a constellation of stars as Leo; Muzak versions of "Stairway to Heaven" or "Black Dog" still qualify as Zeppelin tunes (well, kind of); and the phrase "surf the web" shows that the meaning of that verb is not even yoked to waves or water.

Or to take a slightly different perspective:

- Sensory and motor systems are so tightly interwoven that it makes little sense to view them as the segregated buns of input and output separated by a cognitive meat patty.
- Yet, we can't point to any particular motor representation or any particular sensory representation that captures the scope of conceptual knowledge.

The hierarchical and dual-stream principles together relieve the tension by allowing us to have it both ways. In the dorsal stream, sensory and motor systems are indeed tightly connected by necessity and once called into action carry out their computations *largely* automatically, without substantial intervention by higher-cognitive realms. By this I mean that we don't have to reason through what contraction/

relaxation patterns we have to generate in the muscles of our arms and torso to reach for a coffee cup. Once we decide to reach, the sensory target features of shape, size, orientation, and distance together with proprioceptive information about body position can be transformed automatically into the execution of a reach. There is no cognitive sandwich in the dorsal stream; sensory and motor processes function as a unified system.

However, to make the decision to reach for the cup in the first place, the system needs to do something different. It needs to recognize that the object is a cup and not a saltshaker, that its context in space and time indicates that coffee is inside (even if you can't see the liquid), and that the cup is a socially acceptable target for reaching (it's not someone else's), and all of this information needs to be related to internal states such as a current desire for coffee. To perform this set of operations requires access to more abstract, stable knowledge concerning what coffee cups look like, what coffee tastes like, whether drinking it results in pleasure or just the jitters, what social conventions apply to grabbing coffee cups, whether it is likely to be too hot, too cold, or just right given how long it's been in the cup, and so on. That is, the potential target for action must be interpreted and understood—the function of the ventral stream. Here we have somewhat of a cognitive sandwich: sensory information is processed and related to stored knowledge relevant to the present situation; a decision is made to act on this information; and the (sensori)motor system is called into action, or not.

We can think of these two systems as different levels in the sensorimotor hierarchy of the nervous system as a whole. The dorsal stream controls action at a lower, more direct level. More like a reflex, it doesn't care whether the reach is directed at a cup or a saltshaker—that's not its job—all it cares about is executing a smooth, successful action directed at its target. Also like a reflex, this is a largely unconscious system. If the conditions of the action change suddenly—if the target is moved or if your arm is bumped—the dorsal stream system automatically adjusts the movement trajectory with an effortlessness that is hard to reproduce in robot arms.[17] This is not to say that the dorsal stream is completely impervious to influence. Just like a low-level pain reflex can be overridden by higher-level control, the reflex-like action of skilled movements controlled by the dorsal stream can be meddled with by higher-level "cognitive" systems. This meddling is precisely what we are trying to avoid when we say about our golf swing, "I'm thinking about it too much" or when our trainer advises "just let it go, your muscles know what to do."

The ventral stream, on the other hand, controls action (or not) at a higher, more indirect level. Rather than just reflexively acting on a target, this system *evaluates* the affordances for actions, to use James Gibson's term, using long-term memories of past experiences to categorize and recognize objects, integrate them into an environmental context and our internal state, and then uses this information to make decisions on whether to act now, later, or never. This system allows our incredible flexibility of behavior.[18] Even though we may desperately want that cup of coffee, we may withhold reaching for it for any number of reasons: it belongs to someone else, a fresher one is on its way, it will cause a stomachache, etc. This system is more accessible to conscious reflection and therefore rather more plodding, less automatic.

So the ability to produce adaptable and complex movements requires two kinds of sensory systems. One that registers perceptual features in the environment that are needed for guiding action *now*, and another that evaluates opportunities for action in light of the current state and past history (memory) of the organism and then makes decisions on how or whether to act.

This dual architecture explains why there is plenty of evidence that sensory and motor systems are so tightly interwoven as to be nearly inseparable with little "cognitive" mediation between; this is precisely the situation in the dorsal stream. It also explains why, on the other hand, there is evidence that conceptual knowledge cuts across and goes beyond sensory and motor particulars; because abstraction and integration with other information sources is what the ventral stream does.

We still haven't explained why fairly low-level sensory or motor systems seem to play a role in at least some aspects of conceptual knowledge, why auditory cortex activates when you think about a song, and so on. Put differently, why isn't all conceptual knowledge as abstractly represented outside of sensory and motor modalities? The answer has to do with the hierarchical organization of knowledge and the neural systems that support it.

If I showed you a picture of a cat and asked you to tell me what it is, you would most likely say "a cat." Cognitive scientists refer to this as the *basic category* for the object because it is the most common, or default, name that people come up with to describe it. But a cat is also a member of increasingly abstract categories such as PET, FELINE, ANIMAL, and LIVING THING. We can also zoom in and conceptualize the cat as a house cat, an Abyssinian, or a cartoon Felix the Cat. Importantly, depending on how we conceptualize the object, its conceptual associates vary. As a pet, our cat is associated with dogs, birds, and hamsters. As a feline it hangs with lions and tigers and ocelots. This is relevant because these conceptual relations impact behavior; we make inferences and reason using category membership information. If you know something about one cat—for example, that it crouches just before pouncing on its prey—it is a reasonably safe bet that it applies to other cats. It's a cognitive shortcut to understanding and predicting events in the world: even if you have never seen a tiger before, your experience with house cats is useful in predicting the jungle cat's behavior. Similarly, if you have experience with pets, you can generalize that knowledge to a new pet you might encounter, for example, that it is likely to be *relatively* tame, even if that new pet happens to be a tiger.

If categories are hierarchically organized and if categorical information is important for making cognitive inferences and reasoning, then this implies that the conceptual knowledge system in our brain is hierarchically organized and multifaceted. There is no single conceptual representation of CAT that is completely abstract (the classical view) or completely yoked to perceptual or motor features (the embodied view). Rather, there is more than one way to think about a cat and this is reflected in the different levels of, and associations between, conceptual networks that our cat knowledge is tied into (Abyssinians, pets, felines, animals, cartoon characters, living things). Depending on the context of the cognitive task at hand—deciding what to do if we run into a tiger in the jungle or one on a leash at the circus—we tap into different levels or portions of those knowledge networks to make our inferences and reason through the problem.

So far, so good. But the conceptual levels we are talking about are still fairly abstract, even for Felix who can be realized in various cartoon forms or character costumes. How do we explain why low-level sensory or motor regions get involved sometimes when we think about songs or cats? Picture a cat in your mind's eye. No, not any old, generic cat, but a specific cat in a particular pose. How many whiskers does it have? Given its particular pose in your mind's eye, can you see all of its whiskers or are some obscured? Of those that you can see, what angle is the top whisker relative to the floor? When we think about things in our world we can think abstractly in terms of categories (CATS, PETS, ANIMALS) or we can think about specific details. Some of these details, if they were being analyzed perceptually, would involve fairly low levels of processing.

Now we have the ingredients to explain both the involvement of low-level sensory and motor systems in "cognition" as well as the fact that humans form abstract categories. These ingredients are (1) the hierarchical organization of conceptual knowledge, which can involve anything from low-level perceptual features to very abstract categories, and (2) the hierarchical organization of perceptual and motor systems in the brain, which code everything from low-level perceptual features (lines and edges) to fairly abstract categories (FACES). If we assume that perceptual and motor systems are part of the conceptual system, as the embodied theorists argue, then we have merged the two ingredients and built a hybrid model of conceptual organization in the brain that includes both low-level sensory and motor systems as well as abstract categories. We involve low-level representations when we are thinking about particular details and we involve more abstract, higher-level representations when we are thinking more abstractly. The fact that the hand area of motor cortex activates when we think about grasping a cup doesn't mean that the concept of GRASP more broadly is coded in primary motor cortex; all it means is that we can think about (mentally imagine) the details of a particular grasp. The more abstract concept GRASP, which includes lots of motor possibilities, lots of potential graspers, and lots of objects or even ideas is likely coded at a much higher level and in a system that doesn't have to worry about the particulars of controlling a specific instance of grasping.

Empirical support for a hybrid model of conceptual knowledge

The data back up this hybrid, hierarchical model of conceptual representation. We've already discussed plenty of evidence that lower-level sensorimotor regions activate when we think about particulars. Is there also evidence for a more abstract system that isn't tied to any particular sensorimotor modality? Absolutely. A herculean review of the imaging literature on the "semantic system" makes the point quite convincingly.

A team led by neurologist Jeffrey Binder at the Medical College of Wisconsin individually screened the abstracts of over 2800 published scientific papers reporting functional imaging studies (fMRI and PET) of conceptual processing and fully reviewed 520 of them.[19] They whittled the set down to 120 that fit their criteria for inclusion in a more detailed, meta-analysis. Their criteria were designed to factor out the influence on brain activation patterns of overall task difficulty and decision-making processes, and to ensure that the studies implemented proper controls for irrelevant low-level sensory and motor processing. For example, if a study measured brain activity for making semantic judgments about pictures (e.g., press one button if a picture is a living thing and another if it is an artifact) and compared the activation pattern to a baseline "rest" condition (lying calmly in the scanner and "doing nothing"), it is poorly controlled. The semantic task and the baseline condition differ not only in terms of semantic processing but also in terms of sensory stimulation (seeing pictures) and motor execution (pressing buttons), factors that have little to do with thinking about the meaning of items. Visual and motor cortices undoubtedly activate in such a study even if they don't participate in semantic processing per se. Binder and his colleagues excluded such studies. What the team *did* include were studies that contrasted various types of semantic processing such as comparing meaningful versus meaningless stimuli or actions versus objects or animals versus tools.

From these studies they compiled 1135 brain activation locations associated with semantic processing. It is conventional in brain imaging studies to digitally warp each subject's imaged brain into a common, standard space and report activations in terms of *x, y, z* coordinates within that space. This allows researchers to perform meta-analyses in which activation patterns across studies can be pooled and thrown into one giant analysis that statistically parcels out consistently observed activations from those that are spurious. This is exactly what Binder's group did.

When Binder and colleagues did this, they found a network of cortical areas that participate in conceptual semantic processing, as they defined it. Noticeably lacking from this network were the major swaths of tissue that correspond to sensory and motor systems. In fact, if you took a figure of the brain and shaded in all of the regions typically associated with visual, auditory, somatosensory, and motor processing—including not only "primary" but also all of the extended sensory and motor regions—Binder et al.'s "semantic system" would fall almost exclusively in the nonshaded areas, that is, in regions often characterized as *heteromodal association cortex*, cortex not tied to a particular sensory or motor modality. They conclude that "these observations confirm a general distinction between conceptual and perceptual systems in the brain." They acknowledge that sensory and motor systems may play some subtle role that their analysis could have missed, but it is clear from their work that the bulk of the conceptual semantic system, *as invoked in the tasks they examined,* is not in sensorimotor cortex.

The same point—that conceptual semantic processes can involve abstract representations not tied to particular sensorimotor modalities—can be underlined by another, serendipitously discovered approach. Comedian Steven Wright hit on the basic observation when he said, "I was trying to daydream, but my mind kept wandering." Let me spell out the idea in a brief digression.

It used to be thought, by some at least, that the brain could truly rest, muscle-like, only becoming active when it was roused by some task.[20] When you look at brain physiology, you know that this isn't the case. Neurons do not go silent when not in "use," they just fire at a baseline rate that is then modulated up or down under various "active" conditions. Brainwaves do not straight-line when subjects are told to relax and rest; they continue to oscillate and twitter. Just try this to convince yourself that your brain never really rests: for the next 30 seconds, think about nothing. It's remarkably hard, maybe impossible, to do. Even meditative states—which require *training*—involve a focus of attention on something, be it an object, one's own breathing, a mantra, or an inner, unbounded awareness. This requires physiological work in the brain that is evident in EEG traces and PET/fMRI activation patterns.[21] A recent fMRI study on meditation noted that "long-term practitioners . . . had significantly more consistent and sustained activation . . . during meditation . . . in comparison to short-term practitioners."[22] Thinking about "nothing" may amount to thinking about something in a very focused way.

The brain's restlessness is a problem for many brain imaging studies because we often compare activity associated with the task or stimulus of interest with "rest," a condition in which we ask our participants to do "nothing" and "clear their minds." It doesn't work. As data from functional imaging studies started rolling in in the 1980s and 1990s a consistent phenomenon was observed across many studies of various sorts: a common network of brain areas was more active during the "rest" period than the various task or stimulus conditions, which were often sensory or motor. These *deactivations*, as they were called, were typically discounted by researchers (myself included!) because our focus was on the brain regions that *increased* activity during the task/stimulus conditions relative to baseline, not on the regions that *decreased* activity (that is, were more active during baseline). But some researchers wondered whether we were missing something important.

Jeffrey Binder was among the first to worry about what the deactivations reflected. He came across previous psychological work that indicated that an individual's stream of consciousness tends to wander when the external environment is constant and predictable (i.e., when we're bored). When given a particular task or novel stimulus, however, this mental wandering is suppressed. Binder suspected that the brain activation patterns during "rest" reflected a wandering mind and did an experiment to confirm the hunch. Binder and his team showed that the network of brain areas that activated during "rest" relative to some active task was essentially the same network that activated during a conceptual-semantic task compared to a nonconceptual semantic task. In other words, the "resting" brain is not resting, but actively processing meaningful information, daydreaming, *thinking.*[23]

This phenomenon has become a fairly intense topic of study in cognitive neuroscience and the network underlying it has been termed the *default network.*[24] Disagreement remains about what exactly it is—whether it is one network or many, whether it is best characterized as an autobiographical memory system, a semantic system, some sort of monitoring system, or the neural basis of theory of mind (our ability to recognize that others have minds of their own)—but it is quite clear that whatever it is, it is fairly high level, turns on when not much is happening in the external environment, turns off when focused attention to that external environment is needed, and occupies a large fraction of the territory of the cerebral cortex.[25]

What's the point of this digression? We have a brain network that activates "when individuals are engaged in internally focused tasks including autobiographical memory retrieval, envisioning the future, and conceiving the perspectives of others," to borrow words from Harvard psychologist Randy Buckner and colleagues.[26] By all accounts, this kind of thinking, no matter what you call it, is involved in processing meaningful information about one's experiences. Indeed, the same network is hit particularly hard by the neural degeneration that occurs in Alzheimer's disease, a disorder that dramatically impacts the ability to recall and think about life experiences, events, names, places, words, and the like—all the things that give meaning to life.[27] Crucially, for our purposes, this network specifically does *not* include the systems that respond during immediate interaction with the external environment. It appears that our knowledge of the world and our personal experiences in it rely on the integrity of a high-level network that is largely distinct from our sensory and motor systems. This doesn't mean that useful information isn't or can't be stored in lower-level sensory and motor systems; but it does hammer home the point that a good chunk of our mental world involves fairly high-level cortical systems that are not part of particular sensory and motor modalities.

The broad point, then, is that conceptual knowledge is hierarchically organized. We have more specific sensory and motor knowledge about the lower-level details of our experience encoded in similarly low levels of the cortical hierarchy and we have more general knowledge about the world stored in higher-order cortical systems, and presumably at a range of levels in between. We can then access different levels of this hierarchical network depending on the task at hand. Highly abstract and grounded conceptual representations can coexist.

TOWARD A NEURAL BASIS OF ACTION UNDERSTANDING

WITH THIS broad picture in mind we can sketch an alternative, non–mirror-neuron-centric, view of action understanding. Let's start by considering the neural basis of our knowledge of objects (e.g., FRUITS, TOOLS, ANIMALS), which falls somewhere between low-level properties and über-abstract conceptual knowledge. As such, object concepts serve as a nice model for action concepts.

Decades of research on the neural foundation of object concepts has led to somewhat of a consensus among researchers: object concepts live in the temporal lobes.[28] We know this because damage to this region causes deficits in recognizing objects. Exactly how the system is organized within the temporal lobes is still a matter of debate. A central issue is how to explain the fact, noted in the previous chapter, that deficits can be selective to one category of object or another (e.g., ANIMALS versus FRUITS & VEGETABLES). Some authors, such as the National Institutes of Health neuroscientist Alex Martin, argue that the system is organized according to perceptual features (shape, color, motion), whereas others, such as Harvard University's Alfonso

Caramazza, argue that it is organized by functional category, particularly those with evolutionary significance (ANIMALS, PLANTS, FACES, and maybe TOOLS).[29] This debate is not particularly relevant here. There is also debate about where in the temporal lobe the relevant networks live. Data from functional imaging studies implicate posterior temporal lobe regions, more toward the back of the lobes. Just about any time you ask group participants in a brain imaging study to perform a conceptual task involving objects or names of objects, you find activations involving the posterior middle temporal gyrus.[30] Conversely, data from another source, the degenerative condition known as semantic dementia, have implicated more anterior temporal lobe regions toward the front of the temporal lobes.[31]

Semantic dementia is a scientifically fascinating, if personally devastating neurological disorder. It is a subtype of a class of so-called primary progressive aphasias, Alzheimer's-like degenerative disorders that attack language and conceptual systems particularly.[32] The first symptoms in the semantic variant are difficulty naming objects and comprehending single words, hence the link with aphasia. As the disease progresses, however, patients seem to lose the very concepts of objects themselves.

In a foundational report published in 2007,[33] neurologist John Hodges and psychologist Karalyn Patterson illustrate the degeneration of word knowledge through patients' naming responses over a series of visits to the lab. At first, picture naming errors are semantically in the ball park and at the same level, compared to the target: "giraffe" for zebra, "swan" for ostrich, "bird" for ant, "boot" for sock. As the disease progresses, errors tend to shift toward more common names in the same category or to more basic category terms: zebra is "horse," ostrich is "bird," ant becomes "animal." In advanced stages, "thing" or "something" or "don't know" dominate the patients' responses. And it's not just a word or picture problem. If asked to copy a drawing of a peacock, they can render an easily recognizable copy. But asked to redraw it after a delay of 10 seconds the beast morphs into a different animal, with four legs and a bushy tail, just like the naming responses morph one object (zebra) into another, often more common one (horse). Asked to repeat a complex word like *hippopotamus*, subjects can do it perfectly. But ask what a hippopotamus is and the response might be, "I think I've heard of a hippopotamus but I can't say what it is." It's as if the concepts simply fade over time into a neural fog.

Neural degeneration in semantic dementia can be tracked using imaging methods that assess changes in atrophy of the cerebral cortex, blood flow, and connectivity. All of these metrics point to anterior portions of the temporal lobes, regions quite downstream from early sensory processing, and nowhere near motor cortex.[34] And this is the relevant point. While debate continues about the details of the system within the temporal lobes (the ventral stream), most investigators agree that conceptual knowledge of objects involves moderately high-level systems and does not involve the motor system. You don't get semantic dementia from damage to primary visual, auditory, somatosensory, or motor cortex, that is, the lowest levels of sensory or motor processing cortex. There is *no* evidence that damage to these low-level cortical systems produces any kind of broad, object-related conceptual deficits. Instead, when *primary* areas like these are damaged you see focal, low-level deficits: a blind spot in your visual field, difficulty discriminating between similar tones, lack of sensation or weakness in a limb, and so on. Similarly, damage to the dorsal stream, even when high-level motor deficits such as limb apraxia are present, does not result in semantic dementia.

The neuropsychological evidence is clear: the neural basis of conceptual knowledge for objects critically involves the temporal lobes, not the motor system.

What about action concepts such as GRASP, KICK, and THROW? Mirror neuron or motor embodiment theorists hold that action concepts involve the mirror system, Broca's area in particular, or primary motor cortex itself. The evidence doesn't support this view. Instead, the data are increasingly pointing to the temporal lobes as a critical location for action concept knowledge. Although one study reported an association between damage in Broca's area and the ability to recognize actions such as guitar strumming in a large group of subjects, it was clear from that same study that the correlation was far from causative, as we discussed in Chapter 4.[35] Subsequent studies reported a different pattern. One experiment involving a group of patients with primary progressive aphasia (a different subtype than semantic dementia) found an association between gesture recognition deficits and cortical degeneration in the superior temporal gyrus and a posterior parietal region, not in Broca's area.[36] And a study of 43 stroke patients found a strong association between damage in the posterior middle temporal gyrus and the ability to recognize the meaning of manual gestures, such as hammering.[37] Moreover, although object concepts have been a major focus of research on semantic dementia, the deficit in that syndrome extends to action concepts as well, further implicating ventral stream regions in action recognition.[38]

To be fair, a few published studies appear to directly support the view that the mirror/motor system is at the core of action concepts. One of the first was published in 2003.[39] No fewer than 90 individuals with left- or right-hemisphere brain lesions were studied. Knowledge of actions was assessed in each case (see below for details) and deficits were related to lesion location. The map of the culprit lesions looked a lot like a map of the human mirror system, at least on first glance: all of Broca's area was implicated along with portions of motor cortex and the anterior, inferior parietal lobe—the home of the dorsal, sensorimotor stream. But there was one more lesion location implicated, the same posterior middle temporal region found in more recent studies.

Why might the 2003 study have found such prominent involvement of frontal and dorsal stream regions? It is likely a task issue, similar to that discussed in the context of speech processing in Chapter 5. Patients in the 2003 study were never shown full-blown video-recorded actions, but rather static pictures of actions, a banana or apple being peeled, cards being dealt or shuffled. They were not asked to identify the action (peeling or dealing) but instead had to make inferences about them. For example, if shown a picture of cards being dealt alongside a picture of cards being shuffled, subjects were asked to select the pictured action that would make more noise. Or, in another task, participants were shown three action pictures (such as peeling a banana, peeling an apple, and lifting a lid off a pot) and had to choose the one that didn't belong. Tasks of this sort involve cognitive operations beyond the ability to represent the meaning of the actions, including inferring the action from a static image, inferring properties that are associated with the action such as what noises might result, and comparing associated properties of different pictured actions on some dimension such as loudness. High-level cognitive operations such as these—making inferences, holding information in "working memory" while comparing properties across a set of items, choosing between similar response options—are sometimes called "executive functions," long known to involve frontal lobe brain areas.

It's likely, then, that deficits on the action-inference tasks (which action makes more noise?) could come from at least two sources, one from a deficit in processing action knowledge itself and another from a deficit in performing the executive functions demanded by the tasks. Given that the posterior temporal lobe location has shown up in a majority of studies, it is reasonable to conclude that this region is a critical node in the conceptual network and that the frontal- and motor-related regions are more task dependent. A figure from the 2003 study, which I've reproduced in rough form here, reinforces this conclusion.

These are depictions of the damaged area in two different patients, one with a massive lesion affecting frontal motor areas and the other with a small lesion affecting the posterior middle temporal gyrus. The individual with the larger, motor-related lesion was impaired on one of the two action knowledge tasks (the authors didn't say which), suggesting that under some conditions at least access to action knowledge is preserved, whereas the subject with the smaller

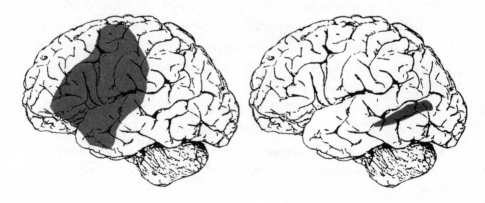

posterior temporal lesion was impaired on both, suggesting a more severe conceptual deficit.

Why might the motor system get involved in "executive function" tasks? Often, these tasks involve some form of working memory, the ability to keep information active and accessible while problem solving. The motor system supports working memory by functioning as a mental scratch pad. If you hear a sequence of numbers that you need to remember until you can dial them on your phone, what do you do? You say them to yourself, which is to say that you recode the numbers in your motor speech system and cycle through the code until you are ready to dial. You don't need to say them to yourself to understand the numbers—you could listen to a set of 50 numbers and report after hearing the whole list whether they were mostly small or large numbers—you just need to recruit your motor system to hold on to the phone number's particular sequence for more than a couple of seconds.[40] Decades of research on working memory in humans and monkeys has shown that the frontal lobe gets involved in all sorts of situations that require temporary maintenance of information, not just remembering phone numbers.[41] But in humans, speech is a very convenient scratch pad for all kinds of information and so we have to be particularly careful to rule out motor speech recoding as a possible source of assistance in any task. In fact, this may be a possible explanation for why Broca's area, which is involved in motor speech, activates in so many tasks that don't seem to require speech production: subjects may just be talking to themselves in their head while in the scanner!

The weight of the evidence points to a model of action under-standing in which the posterior temporal lobe, rather than the motor system, plays a central role. I don't mean to imply that this region is a phrenological island for the storage of action knowledge. It is more likely a kind of hub, or *convergence zone,* to use the terminology of neurologist Antonio Damasio, that binds together a broader network that represents and processes information related to actions.[42]

This broader network likely includes a nearby area known as MT (for middle temporal) that is long known to be involved in visual motion processing generally in both monkeys and humans. The broader network also likely includes links to more specific knowledge about how *bodies* tend to move, *biological motion perception*. Humans are remarkably good at identifying biological motion, to the extent that we can do it based on very little visual information. Swedish psychol-ogist Gunner Johansson demonstrated this convincingly in a paper he published in 1973. He reported a study in which he placed 10 patches of highly reflective tape on the major joints of an actor–assistant and then videotaped the actor walking in the flood of a bright light. When the contrast of the recording was cranked all the way up, only the joint reflections were visible, resulting in what are termed *point-light anima-tions*. Johansson then showed the animations to observers and asked what they saw. "It always evokes the same spontaneous response after the first one or two steps," he wrote, "this is a walking human being!" (Demos are readily available. Just Google "point-light walker.") Sub-sequent work has shown that many actions are as easily and immedi-ately recognizable. It's a remarkable ability. As Johansson put it, "How can 10 points moving simultaneously on a screen in a rather irregular way give such a vivid and definite impression of human walking?"

We don't know the answer to this question entirely, but neurosci-ence research using point-light animations has identified the critical brain region: the posterior superior temporal sulcus (STS; see figure in Chapter 3). In functional imaging studies, it responds more to recog-nizable point-light animations than to random movements of the same lights. Further, zapping the region with TMS disrupts subjects' ability to recognize the movements in the displays.[43] The same region activates quite well to plain old, fully visible human actions, including reaching and grasping actions, lip movements, and shifts in eye gaze.[44] Recent work has shown that activity in the posterior STS is exquisitely sensitive to socially relevant actions. It differentiates between expected and unexpected shifts in observed eye gaze (looking toward versus away from an object that appeared in front of the actor); it differentiates between gaze toward the observer versus away; and it differentiates between the observation of actions that achieved the intended goal versus actions that failed (attempting to place a ring on a post but missing).[45] Findings such as these have led to the proposal that the posterior STS plays "an important role in detecting, predicting, and reasoning about social actions and the intentions underlying actions"[46]—exactly the kind of function that mirror neurons are supposed to explain.

Very similar kinds of responses have been found in single neuron recording studies in the monkey STS. Monkey STS neurons respond to actions such as walking toward or away, head turning, movement into or out of view, arm movements, and hand–object interaction where there is selectivity for specific actions including reaching, retrieving, manipulating, picking, tearing, presenting objects to the monkey, and holding.[47] Monkey STS cells are also sensitive to eye gaze direction and head orientation of the observed.[48] And more broadly, monkey STS also responds to point-light displays of the actions of other monkeys.[49]

This is a rather impressive constellation of response properties for a region that according to the mirror neuron theorists is not involved in action understanding. I suggest that we take a more straightforward theoretical position and hypothesize that, given the sensitivity of the STS to a wide range of actions, including those that are highly socially relevant, the STS is an important, indeed critical, node in the action understanding network. The mirror system is not the only game in town when it comes to identifying the neural basis of action understanding in monkeys and humans.

Other systems that are likely linked up to the posterior temporal (and perhaps STS) action understanding convergence zone include temporal lobe regions that could put the perceived movement, say, reaching for a cup, in the context of (i) surrounding visually perceived objects (e.g., a café table, chairs, plates, utensils; the lateral occipital cortex), (ii) higher-order semantic representations of objects (e.g., what cups and plates generally look like and what they are used for; anterior and/or posterior temporal lobe), (iii) the particular people involved (e.g., a friend, spouse, colleague; the fusiform face area), (iv) the broader semantic or autobiographical context (e.g., what's the purpose of the event; the default network and medial temporal lobe memory areas), and (v) the emotional relevance (e.g., friendly or confrontational encounter; amygdala), which all together put the meat of understanding on the bones of the perceived action. The network also has to be interfaced with prefrontal circuits involved in (i) assessing the relevance of the action given the context above, (ii) directing attention to the action if the context dictates, and (iii) selecting an appropriate response, if any. Further, this network is surely linked up to language systems for thinking of the name of the action, to auditory systems for acoustic associations, to somatosensory systems for touch associations, and to motor systems for imitating or for otherwise reacting to the action, should a response be warranted.

Now isn't *that* a complicated, unruly network! One might counter that this is no less than a course model of the function of the *entire* brain from the perspective of some hypothesized core "node" involved in performing laboratory tasks related to viewing actions. Surely it's too complicated and unparsimonious to be a useful model of how the brain understands actions.

Well, that's kind of the point. "Understanding" is a complicated thing with lots of moving parts. You can't pull out one part and call *that* the "basis of action understanding." It doesn't work for just the motor bits and it doesn't work for just the sensory bits. Nor is the one high-level posterior temporal STS and/or MTG bit enough to define what it means to understand, say, GRASPING in all its guises.

Action understanding is the interaction of many things. At the core, arguably, is the action itself and the neural networks involved in processing those movements; likely the STS is critical to this part of the system. But the movements alone are not enough—on this I agree with Rizzolatti and colleagues—they must be linked with the context of the action. And the interpretation of this information must be further influenced by the observer's knowledge of the people or critters involved, their personality characteristics, emotional state, and their past actions in similar situations. The observer's cognitive and emotional state also interacts, of course, with the action itself and other factors in arriving at a full understanding (or misunderstanding—we sometimes get it wrong).

In the context of a model like this, to say that we understand action by simulating that action in our own motor system doesn't make much sense. The motor part, while associated with the action for sure, is such a minute part of the bigger picture as to be virtually useless on its own. At the very most, one could possibly argue that accessing a motor copy of the perceived action could help facilitate or speed access to the broader, nonmotor network that enables understanding.[50] But this is far from the claimed *"basis* of action understanding" that generated the excitement about mirror neurons in the first place. And moreover, why not just access the broader network from the sensory representation, which your brain has to engage anyway and which, by all accounts, is also associated with the rest of the network? If we rely on the lesion data as a guide, it seems that the perceptual system, such as the STS convergence zone, is the real gateway to understanding, not the motor system.

If the whole brain is involved in understanding, isn't this just a restatement of the embodied cognition view? Well, if by "embodied cognition" one means that the brain is a computational network at all levels, that it is hierarchically organized, that knowledge comes in different degrees of granularity reflected by the neural hierarchy, and that knowledge is organized around and accessed according to the particular task an organism must perform, then sure, it's embodied. If by embodied cognition you mean that understanding is only the result of low-level sensory or motor (re-)activation, without abstraction, without information processing, and without a differentiation of systems according to task goals, then no, it's not embodied. In the end, it doesn't matter what we call it. The goal here is to explain how the brain works, not to name the philosophical approach.

Yet another possible objection is that the mirror neuron theory of action understanding was never intended to explain the entire process, just the part that allows us to distinguish between a grasp and a flick. Just because understanding actions at their deepest and most nuanced levels might involve brain systems outside of the motor system doesn't mean that motor simulation isn't at the core of a more basic form of action recognition. Fair enough. But as we've seen, if we restrict our focus to just this more basic form, the evidence points to a hub in the posterior temporal lobe rather than the mirror system. And more broadly, it is not enough to understand that someone is grasping. To make sense of the world on a level that allows us to act in a purposeful way we need to understand the causal forces behind the actions. By understanding that grasping a menu is caused by the desire to choose a dish, which is caused by a feeling of hunger, we can not only understand the cause of that particular action but also generalize the causal force to explain or predict other actions. For instance, we might be able to recognize that a radically different action, like visually scanning the restaurant for a specials board, should be interpreted exactly the same as the grasping action. This kind of understanding cannot be handled by the motor system or the sensory system alone— because there is nothing motorically or sensorily in common between grasping and visually scanning—but rather requires a much deeper basis for interpretation.

The alternative view of the neural basis of action understanding sketched here is just that, a sketch. It doesn't explain action understanding any more than "motor simulation" explains it. For a real explanation we need explicit computational theories that detail how neural firing patterns within complex neural architectures enable the brain to transform a visual pattern of form and motion on the retina into a representation of a recognizable movement pattern and how such a representation is linked via additional and equally complex neural networks to the spatial and temporal context of the action, to past memories, to emotional states of the observer and the actor, and so on. And it requires a computational theory of how this network of information is integrated and transformed into a representation that reflects a level of understanding and an action plan (or nonaction plan) appropriate to the observer's task at that particular moment. The sketch doesn't come close to this. What it does, though, is provide the framework for fleshing out some of the components. And lots of scientists are already working on this problem, even if they don't know it, with much success. Visual scientists are making excellent progress in understanding how form and motion and even biological motion information are processed; motor control researchers are working out the circuits that control action; other researchers are uncovering the details of how information from different senses is integrated cross-modally; there is a ton of work on how memories are laid down and accessed; and a good number of brave souls are venturing into the prefrontal cortex to work out how information is integrated and decisions made, and how emotional information plays into these processes. We have a sketch of the puzzle shape and many of the pieces. We just have to sit down and work out the details of how they all fit together, and then see what else we're missing. Some researchers, such as Michael Arbib and his collaborators, are already working toward this.[51]

So, if mirror neurons aren't the basis of action understanding, what are they doing? We take up that question in the next chapter.

8

Homo Imitans and the
Function of Mirror Neurons

It is widely celebrated that our species manifests creativity and thought (Homo sapiens). Evidence for this is coupled, especially during our early period of rapid psychological growth, with a powerful complementary proclivity for re-creation and imitation (Homo imitans).

—ANDREW MELTZOFF, 1988

HUMAN COPYCATS

IMITATION MAY be the sincerest form of flattery, but the compliment is rarely taken well. We admire social trendsetters and political leaders, while we scold our kids for following the (wrong) crowd. We reward entrepreneurial innovators in design and technology, while we take the copycats to court for patent infringement. We give lofty awards to individuals with novel ideas and unique accomplishments in the arts and sciences, while we devalue derivative efforts and punish plagiarism. Imitation, whether it's wannabe crab or a cubic zirconium, is rarely valued.

But if developmental psychologist Andrew Meltzoff is right, we can't help but copy. It's part of being human and it's prewired into our brains. Moreover, imitation may be a uniquely human ability and a key ingredient in the recipe for sapience. Humans ape better than apes ape and this, supposedly, gives us an intellectual leg up.

Meltzoff came to this conclusion based on a series of experiments carried out over the last three decades that documented the human infant's impressive capacity to imitate. The most jaw-dropping of these studies involved testing newborns while they were still in the hospital maternity ward; the youngest participant was forty-two minutes old. The experiment involved modeling one of two actions for the newborns, opening the mouth and sticking out the tongue. Infants tended to stick out their tongues more during blocks of time when the experimenter was modeling the tongue gesture and tended to open their mouths more during mouth gesture blocks.[1] Subsequent experiments by Meltzoff and other researchers with older infants provided evidence for object use imitation (learning how to use an object by watching),[2] speech sound imitation,[3] imitation from memory (delayed imitation),[4] the ability of the infant to distinguish purposeful from accidental actions in their imitation (they don't imitate accidental actions),[5] and the capacity to imitate the *intention* of an observed action rather than the *actual outcome* in cases where they differ (trying but failing to achieve some goal).[6]

So imitation comes in early, it's widespread, and it's smart. What purpose does it serve? If recent theorizing is correct, its purposes include social learning (learning by observation), cultural transmission, the development of theory of mind, and empathy.[7] Let's break down the arguments.

Social learning is probably the most obvious purpose of imitation. If a child can learn from an expert model—from someone who has already solved the problem—learning can be dramatically accelerated. There's good evidence that youngsters do this quite readily. One study employed "expert teachers"—fourteen-month-old infants who had already learned how to operate novel toys—to model actions for fourteen-month-old "naïve infants" who did not know how to work the toys. The naïve infants were allowed to observe the teacher-infants operating the toys, making them open, collapsing them, making them buzz, and so on. Then the naïve infants were removed from the lab/playroom environment. Two days later, new members of the experimental team visited the naïve infants' homes and presented the toys. The infants correctly operated the toys more successfully than control infants who did not have the actions modeled for them.[8] Little children indeed learn from watching their peers, as many parents know all too well.

Humans have institutionalized this kind of learning, of course, in the form of schools, training programs, intern- and apprenticeships. The phenomenon permeates everyday language we use with our children, sometimes encouraging imitation while other times discouraging it: *Follow my lead. Do as I say, not as I do. Watch and learn. Think for yourself! Don't reinvent the wheel.* You can see the proclivity for youthful imitative behavior, too, in the wide array of play gadgets designed explicitly for imitation of modern human culture: toy cell phones, cook sets, tools, dolls, motorized cars, laptops, doctor kits, golf clubs, you name it.

This kind of social learning is nothing new in human history. A Neolithic era archeological site in southern Sweden even turned up some evidence for imitative flint knapping play. (Flint knapping is the process of shaping a piece of flint into a stone tool by striking it with another rock or hard object.) One small region at the site showed evidence of skilled, systematic production of a specialized axe tool, while a nearby and somewhat more distributed region reflected the unsystematic production of flakes, as if a youngster play knapped in a restless, wandering fashion alongside the expert toolmaker.[9]

Whether this interpretation of the archeological site is correct, it illustrates the *cultural transmission function* of imitation learning, which is to pass down technologies and other cultural memes from generation to generation. If we can learn from an expert how to knap an axe from flint or remove the bitter and potentially toxic tannins from acorns, we don't have to figure it out ourselves, which confers survival value and frees the learner to spend time developing new technologies. In short, social learning allows culture and technology to *accumulate* in our species.

There is little disagreement over the importance of imitation-based social learning or that humans are exceptional at exploiting their talent for it. A more controversial idea suggests that imitation is an innate foundation for the human capacity for *theory of mind*. Here's the three-part developmental argument.[10]

1. *Humans are born with the ability to recognize (not necessarily consciously) the equivalence between perceived actions in others and executed actions of the self.* This is the *correspondence problem*, as some scientists have termed it, and it's nontrivial.[11] Having never seen her own tongue or even her own face, how does the newborn know that the pink slug sticking out of the hole in the lower third of the oblong object in front of her corresponds to something in her own mouth that can be controlled by activating a particular motor program? Infants seem to be able to imitate simple movements like this at birth; this means that it can't possibly be learned by experience; and therefore genetics must have solved the correspondence problem for them by prewiring their brains to perform the matching between pink slug and the tongue-controlling motor command.

2. *Through experience, infants learn the relation between their own movements and their own mental states.* They learn that their own desires, for example, are correlated with movements that aim to satisfy those desires (reaching for a toy, bringing food to the mouth or pushing it away) and that emotional states are correlated with particular bodily states (hurt = cry, frustrated = tantrum, and so on).

3. *The ability to recognize a correspondence between the self and others' actions (part 1) plus the realization that self-actions are caused by particular mental states (part 2), allows the inference that others' actions are caused by mental states of their own, that is, a theory of mind.*

This is powerful stuff. If correct, it means that imitation is at the core, the very foundation of what it means to be human both culturally and socially. And this is where the story comes back to mirror neurons.

MIRROR NEURONS AND IMITATION

MIRROR NEURONS seem tailor-made to assume the role of the neural mechanism for imitation. After all, they respond not only during action execution but also during the observation of those same actions. If mirror neurons support imitation, they could "ground" human cultural and social evolution in a fairly simple bit of neural machinery. This possibility seems to fit the mirror neuron excitement perfectly: cells that perform a simple function provide the key to understanding the complexities of the human mind through the mechanism of imitation.

But there's a problem. Despite some early claims that mirror neurons serve an imitative function rather than support action understanding,[12] neither the Parma group nor developmental psychologists such as Meltzoff believe that macaque mirror neurons have much to do with imitation. Why? Macaques don't imitate. Here is Rizzolatti's defense:

> Although laymen are often convinced that imitation is a very primitive cognitive function, they are wrong. There is vast agreement among ethologists that imitation, the capacity to learn to do an action from seeing it done . . . , is present among primates, only in humans, and (probably) in apes. . . . Therefore, the primary function of mirror neurons cannot be action imitation.[13]

And Meltzoff's:

> Furthermore, monkeys do not imitate although they certainly have the basic mirror neuron machinery. Something more is needed to prompt and support behavioural imitation, especially

the imitation of novel actions and imitation from memory without the stimulus perceptually present [which are properties of human imitation].[14]

Something more is needed beyond mirror neurons to explain human imitation.[15]

Based on some of his published statements, Rizzolatti seems to disagree about the role of mirror neurons in human imitation: "the mirror neuron system is the system at the basis of imitation in humans."[16] But in fact he is forced by logic to agree, otherwise macaques *would* imitate. This logic seeps out in other statements by Rizzolatti and his coauthors, such as this one from a discussion of the role of mirror neurons in the evolution of language (the emphasis is mine):

The mirror-neuron system in monkeys is constituted of neurons coding object directed actions. A first problem for the mirror-neuron theory of language evolution is to explain how this close[d], object-related system became an open system able to describe actions and objects without directly referring to them. *It is likely that the great leap from a closed system to a communicative mirror system depended upon the evolution of imitation . . . and the related changes of the human mirror-neuron system: the capacity of mirror neurons to respond to pantomimes . . . and to intransitive actions . . . that was absent in monkeys.*[17]

Mirror neurons alone aren't sufficient; something else had to evolve, imitation, to enable mirror neurons to support lofty human behaviors such as language. From this vantage point it would seem that neuroscientists have been barking up the wrong tree. The obsession with mirror neurons has obscured the real basis of human language, theory of mind, and empathy: imitation!

Step back and we see a twisted tale. Let me review the highlights. According to many researchers, imitation is the key to understanding the impressive advances in human social cognition and culture. According to another group of researchers, mirror neurons are the neuroscientific basis of the very same advances. And indeed, there is evidence that imitation tasks activate the presumed human mirror system, Broca's area in particular, perhaps more reliably in fact than observing the object-directed actions that drive the monkey mirror system.[18] (See Chapter 3.) However, all agree that macaque-type mirror neurons alone do not support imitation—that something beyond mirror neurons enabled human imitation—despite the fact that they seem purpose built to do so.

To sharpen the point, consider the following statements, each considered to be true of mirror neurons according to the Parma theorists (my commentary on the "truth statements" appears in brackets):

- Mirror neurons do not support imitation in macaques [even though they seem capable of matching observed with executed actions].
- Mirror neurons in humans also match observed actions with executed actions; humans imitate, but this ability required the evolution of imitation and "related changes" in mirror neurons. [How mirror neurons evolved or changed to support imitation is not specified.]
- The mirroring mechanism—the simple resonance between observed and executed actions, supported by mirror neurons [and the same mechanism that isn't enough on its own to enable human imitation]—is the basis of language, empathy, and theory of mind in humans [but doesn't support these abilities in macaques even though they possess the same mirroring mechanism].
- Mirror neurons in both macaques and humans support action understanding [despite the fact that this has never been directly tested in macaques and despite evidence to the contrary in humans].

When you get halfway into a crossword puzzle and you find that the correct answers to a number of different clues are at odds with the ones you've already filled in, it's time to reexamine your earlier answers. Mirror neuron theorizing has reached that point, and the juxtaposition of mirror neurons and imitation ability further exposes this state of affairs.

IT'S TIME TO HIT THE RESET BUTTON

LET'S TRAVEL back in time to 1992, open a copy of *Experimental Brain Research,* and begin reading a report titled, "Understanding Motor Actions: A Neurophysiological Study." After reading through the details of the original experiment that uncovered the existence of mirror neurons (although not yet by that name) we read the very first paragraph of the discussion. It goes like this (citations removed):

> One of the fundamental functions of the premotor cortex is that of retrieving appropriate motor acts in response to sensory stimuli. Evidence has been provided that action retrieval can occur in response to two-dimensional patterns, color, and size and shape of three-dimensional objects. The present data indicate that in addition to these physical factors, retrieval can also occur in response to the meaning of the gestures made by other individuals. If one considers the rich social interactions within a monkey group, the understanding by a monkey of actions performed by other monkeys must be a very important factor in determining action selection. Thus, the capacity of inferior premotor neurons to select actions according to gesture meanings fits well in the conceptual framework of current theory on the functions of premotor cortex and expands it to include movement selection related to interpersonal relations.[19]

Stop! This is a brilliant theory. Premotor cortex is part of a sensorimotor association system that can take as input any number of features—size, shape, color, objects, gestures—and use this information to select an appropriate action. This doesn't mean that these cells are responsible for *understanding* size, shape, color, objects, or gestures; it only means that information from these stimuli can be used to guide action selection. The theory "fits well in the conceptual framework" of this sensorimotor circuit in the dorsal stream. All is good.

But mirror neurons ended up being treated differently than other cells in the same region for reasons that we now know: the behavior that mirror neurons seemed most exquisitely tuned to support, imitation, was not observed in macaque monkeys. In fact, in the early 1990s, imitation was widely believed to be a (virtually) unique human ability, "Homo imitans." Given that there was no obvious behavior in the macaque repertoire that a mirror neuron's response could support, it was a natural theoretical move to abandon the action selection theory and pursue other options. Echoes of the motor theory of speech perception still appeared in the literature at the time and so provided the impetus for considering a *perceptual* function for mirror neurons. From there it was a short logical path to theoretical explanations for language, theory of mind, empathy, and, ultimately, a head-on clash with empirical facts to the contrary.

But being time travelers we now have two bits of important information that weren't available in the early 1990s. One is that the action understanding theory would run into factual problems. The other is that macaques *do* imitate.

Research over the last two decades has refined, or more accurately expanded, our understanding of imitation. Psychologists identify at least two types:

> mimicry: sometimes called *simple* or *automatic imitation*, mimcry involves copying actions that are already part of the observer's action repertoire.

> imitation learning: sometimes called *complex imitation*, *true imitation*, or *observational learning*, imitation learning involves copying a sequence of *novel* movements learned from a model.[20]

Moreover, a distinction has also been made between *imitation*, copying body movements, and *emulation*, copying endpoints or goals of actions, such as the movement of objects. There are many potential imitation-like social learning behaviors beyond mimicry that we might explore in connection with mirror neurons in monkeys.

MONKEY SEE, MONKEY DO

WHEN WE look with this wider-angle lens, we find plenty of evidence for imitation in monkeys, as well as in a range of other species. In one study,[21] two macaque monkeys named Horatio and Oberon took turns serving as expert teacher and student on a task that involved touching a set of four random pictures in a predetermined sequence. The pictures were presented simultaneously on a touch-sensitive screen; on each trial the four pictures appeared in random locations on a 16-position (4 × 4) grid display. During training Horatio and Oberon learned their sequences by trial and error; when they hit on the correct sequence they got a reward. They learned a total of 15 sequences with different sets of four pictures, so it was a lot to remember. On average it took Horatio between 19 and 20 tries and Oberon just under 16 tries to hit on the correct sequence with a new set of pictures. After a little practice they could get the learned sequences right with 75 to 80 percent accuracy.

Once they had learned their respective picture sequences, they were presented with the task of learning their experiment mate's list of 15. But, before they started tapping away they were allowed to watch the expert teacher do a few trials. The question was, could the student monkey learn from the teacher and hit on the correct sequence for a new list sooner than he could using trial-and-error learning? The answer was a resounding yes! Both Horatio and Oberon learned by observing the expert. On average, Horatio hit on the correct sequence in about 7.5 fewer tries and Oberson nailed it in about 6 fewer tries.

This shows conclusively that the actions of a model monkey are "a very important factor in determining action selection" in an observer monkey. (A follow-up study using the same task showed that two-year-old children show the same ability to learn by observation and in fact make errors of the same type and at the same rate.)[22]

More evidence for social learning in monkeys emerged from studies of free-ranging troops of Japanese macaques near Kyoto, Japan, who acquired a peculiar, apparently nonadaptive "stone-play" behavior.[23] Stone-play behaviors include gathering and piling, rolling in the hands, clacking, cuddling, pounding, pushing, throwing, and more—45 behaviors so far, to be exact.

The behavior seems to have originated with a single, innovative (or bored) juvenile female in 1979 (the free-ranging troop has been studied continuously for 30 years). Transmission of this stone-play behavior was at first horizontal, passed around among playmates. But when these juveniles grew up and had offspring of their own (around 1984), the behavior was transmitted vertically from elders to younger individuals. Since that time, stone-play has been picked up by every infant in the entire group; but at the time of the first innovation in 1979, it was not acquired by any monkey over the age of five years.

Today the behavior has spread to include four captive and six free-ranging troops in Japan. What's interesting is that the patterns of stone-play observed in the different troops vary regionally, with the degree of similarity of the patterns related to how nearby the troops are to one another. If animals from two troops cross paths frequently, they share more stone-play patterns. If they rarely cross paths, they share far fewer patterns, much like regional dialects. Longitudinal studies of the behavior over a decade and a half have shown that the longer stone-play has been around in the group's experience, the more diverse and complex the behavior becomes, a so-called *ratchet effect*. Controlled studies of captive stone handling macaques have found that the age at which an infant monkey acquires stone-play is predictable from the frequency of stone-play by the infant's mother.

The conclusion drawn from this fascinating set of facts is that these monkeys are passing down a cultural tradition vertically from one generation to the next via social (imitative) learning. Or put differently, the actions of a model monkey are relevant to the selection of actions in the observer monkey.

The observational learning behavior of these Japanese macaques is not a freak occurrence. Similar examples of social, observational learning have now been demonstrated in other species including marmosets,[24] domesticated dogs,[25] mongooses,[26] bottlenose dolphins,[27] bats,[28] fish,[29] and invertebrates.[30] The ability to observe an action and use that visual input to select an action appears to be a common one in animals.

From whence come mirror neurons?

Here were have a potential behavior that mirror neurons might support, not simple imitation (mimicry), but some form of social or imitation-like learning. But what kind of social learning could mirror neurons support in the context of the experimental paradigm that led to their discovery? Psychologist Cecelia Heyes has a simple explanation. Following proposals put forward in the early 2000s by computational neuroscientist Michael Arbib and colleagues,[31] Heyes argues that it is pure association. Macaques reach for and grasp things all the time and they observe their own actions visually. Pretty soon, an association builds between the execution of an action and the (self) observation of that action.[32] Poof! Mirror neurons are born. Now, when the animal sees the experimenter execute an action similar to those that the monkey has previously executed, the cells fire because of the preexisting association built on self-observation. It's got nothing to do with understanding. As evidence for her position, Heyes points to her work in humans, discussed in Chapter 4, showing that changing the association between observation and execution changes the response properties of the mirror system.

I like this idea and I agree that mirror neurons are part of a highly plastic sensorimotor association circuit. But Heyes's idea requires the cells to generalize from the observation of self-action to another's action. This may seem like a trivial step, but as Heyes herself points out, the "correspondence problem"—capturing the relation between self and other action—is not so trivial. This is not an unsolvable problem,[33] but it is a wrinkle.

Here's another thought, hinted at in Chapter 4: recall that during training for the mirror neuron experiments, monkeys observe their human experimenters repeatedly placing and grasping objects as they put new objects into the box and take out old ones. These are the same objects that the monkeys themselves are tasked with reaching for and grasping. This is how mirror neurons were serendipitously discovered in the first place: the monkeys were watching the experimenters grasping objects in between trials and the cells started firing. Furthermore, in some situations the experimenter grasps an item and then presents it to the monkey. For example, this statement is from the 1996 monkey mirror neuron report:[34]

A tray with a piece of food was presented to the monkey, the experimenter made the grasping movement toward the food and then moved the food and the tray toward the monkey who grasped it. (p. 596)

And this is from a 2001 report:[35]

Before starting the neurophysiological experiments, the monkeys were habituated to the experimenters. They were seated in a primate chair and trained to receive food from the experimenters. Monkeys received pieces of food of different size, located in different spatial locations. This pretraining was important for subsequent testing of the neuron's motor properties . . . and for teaching the animal to pay attention to the experimenters. (p. 99)

It appears that the Parma group's training procedures involved exposing the monkeys to a good deal of human actions that are directly related to the monkey's task of reaching for and interacting with objects.

We already know that object shape and size is an important fac-

tor in action selection. Simply seeing a particular object activates
the motor cell involved in selecting an appropriate grip type for that
object; these are canonical neurons. Now, given that the monkeys
are trained to pay attention to the actions of the experimenters, and
given that these actions are relevant cues for action selection (e.g., the
experimenter's particular action or grip type correlates with the shape
and size of the object that the monkey is tasked with grasping), it is
no surprise that motor cells learn to respond to actions themselves. It's
simple classical conditioning.

This may be why the vast majority of mirror neurons are grasping
or placing related, compared to manipulating or holding: grasping
and placing were probably more representative of the monkey's training and these actions are stronger cues for the monkey's own actions.
This may also be why mirror neurons don't respond to grasping with
pliers; because they weren't trained with pliers and therefore couldn't
build up an association . . . *until*, that is, the monkey has a lot of experience with observing experimenters using pliers to place and pick up
objects, at which time mirror responses can be observed, as the Parma
group's 2005 experiment showed.[36] This could also explain why mirror neurons don't respond when actions are not directed at an object
as in pantomime; if there is no object, the monkey has no reason to
select an action. And finally, this may explain why mirror neurons
appear to respond to the goal of an action; because the monkey's own
goal-directed actions are being selected by (have been paired with)
the particular contexts that determine those actions in the first place.
If a monkey is trained to grasp objects and put it in a cup *when the cup
is present*, then the context has been associated with the goal-directed
action. When the monkey then observes the same objects in the same
context, this could activate the associated goal-directed action. Again,
this is classical conditioning.

In Chapter 1 I mentioned, almost in passing, some of the characteristics of mirror neurons reported in the very first paper in 1992 that
did *not* show a congruent relation between their preferred executed
and observed actions. This is worth revisiting. The Parma researchers
write of one subset of such cells:

The effective observed actions were *logically* related to the effective executed actions and could be seen as preparatory to them. For example, the effective observed action was placing an object on the table, whereas the effective executed action was bringing food to the mouth or grasping the object.[37]

Here are the numbers:

39–number of cells that responded to action observation

12–number of congruent mirror neurons

11–number of cells that had a "logical relation" between observation and execution, such as placing $\rightarrow$ grasping

Congruent mirror neurons underpin the whole theoretical enterprise, yet a nearly identical number of cells responded in an antimirror, "logical relation" fashion (coding a relation important for action selection): you place, I grasp. In the larger 1996 report, "*placing* mirror neurons" were the second most common subtype after the *grasping* type, and were found more than twice as often as any other subtype. Had this common class of mirror neurons been singled out for theory development, we would likely have a story not about action understanding but about action selection.

I'm suggesting an associative account of mirror neurons similar to the one Arbib and Heyes promote, but with a different source of the association: the experimental training itself rather than self-action to other-action sensorimotor generalization. The mirror neuron research team may have inadvertently trained mirror neurons into the monkey's brain. Hopefully, future experiments will be designed to test this hypothesis.

Associative accounts of mirror neurons, like the one I've just outlined, have a tremendous additional theoretical appeal. They possess the kind of sensorimotor adaptability that a motor system needs and the empirical record demands. We know that sensory features such as object size, shape, color, location, smell, sound, and so on can be relevant for informing action selection, and we need to be adaptable to changes in these features. Similarly, the actions of other creatures, including conspecifics (animals of the same species), predators, and prey, are highly relevant to action selection. It is critical for an animal's motor system to be able to associate the movements of, say, a snake with appropriate response actions, even though the observer animal may not be able to slither or coil. For some critters, it may be useful to pay attention to the actions of conspecifics and imitate them in one way or another, that is, to mirror them as in cases of social learning, like the stone-play behavior of the Japanese macaque troop near Kyoto or Horatio and Oberon. At the same time, these same critters may find it equally important to pay attention to the actions of conspecifics and select completely different, "antimirror" or "logical relation" actions, such as blocking or fleeing from an attack, or submitting to a grooming gesture. The experiments reported by the Parma group provide empirical support for the existence of such "antimirror" (you-do-this-I-do-that) action association cells.

My broader point is that the Parma team's very first interpretive impulse was correct; the actions of other animals are relevant to the selection of the observer's actions. And now that we know there are plenty of behaviors that mirror neurons could support in macaque monkeys, as well as in other species, there is no theoretical pressure to abandon the idea that mirror neurons support imitation in a broader sense of associations between actions, as in observational learning.

From whence comes homo imitans?

We are left with a problem. If mirror neurons are nothing more than sensorimotor association cells, and if imitation can be conceptualized as a correspondence or association between the actions of self and the actions of others, and if macaques already have this fundamental neural mechanism, then why don't they imitate as prodigiously as we do? And to take the paradox one step further, if imitation is the key to sapience, why aren't macaques (and many other species) as sapient as *Homo sapiens*? *Something more is needed.*

I think the very existence of this paradox points to a logical error in thinking about imitation as the foundation for more complex capacities like theory of mind, or that imitation *itself* had to evolve to unleash a great leap forward. Maybe we should think the other way around. Imitation is not the cause but the consequence of the evolution of human cognitive abilities. Macaques don't imitate on a scale that humans do, not for lack of the foundational neural machinery—they clearly have it in mirror neurons!—but because they don't have the cognitive systems in place to get as much out of imitation as humans do.

Imitation all by itself isn't that useful. Consider language, which on the surface is a prime example of human imitation. After all, we learn the language we hear (or see) as children and end up "imitating" our parents', siblings', and peers' speech sounds, words, and sentence patterns. But we don't exactly. While we do reproduce the phonemes, words, and so on, we don't slavishly imitate pitch, the rate of speech, or the pauses, for example. Nor do we say *dog* while gesturing with our left hand and *lizard* while gesturing with our right because mom is left-handed and dad is a righty. We imitate the relevant parts and largely ignore the rest (but see below). And of course language is more than just imitating what we hear.

Imitation studies in children make the same point. If a youngster watches a model trying to pull apart a novel toy (that in fact pulls apart), but the model's hand repeatedly slips off the end, the kid doesn't imitate the slippage. Instead, she gets a firm grip and pulls the toy apart successfully.[38] Nor would the child imitate the experimenter's coughing during the trial, or scratching her nose, or clearing her throat.

We don't imitate everything all the time, even though we could. Children sometimes do it as a game when they copy or talk along with their playmate's utterances (speech scientists call this *shadowing*). But this isn't what happens in learning situations, and mostly it's annoying.

Further, imitation alone isn't enough. Parrots imitate speech but one parrot can't tell the next that the humans left for the store and will be back in 10 minutes. Even in humans, imitation of speech isn't enough to enable language. Kids never hear their parents say *goed* and *runned* but this doesn't stop the young language learner from over-generalizing (they can't simply be imitating).[39] And infant Japanese macaques imitate stone-play but they aren't crafting axes. *Something more is needed.*

For imitation to be at all useful, you have to know what and when to imitate and you have to have the mental machinery behind the imitative behavior to put it to good use. To fully understand imitation, perhaps we need to look at the mechanisms, the cognitive domains that define the relevant features of the stimulus and therefore define the goals of imitative actions. More specifically, to understand the role of imitation in language learning, we need to study how language works; to understand the role of imitation in the development of theory of mind, we need a model of theory of mind; and so on. Or to frame it a bit differently, rather than centering our theoretical efforts on imitation and then seeing what computational tasks imitation might be useful for, we might center our focus on particular computational tasks (language, understanding actions, grasping for objects) and then see what role imitation may play. This approach, I suggest, will reveal the "something more" that drives imitation.

This view is wholly consistent with Heyes's position that imitation itself isn't a specially designed cognitive adaptation but rather a simple associative mechanism present throughout the animal kingdom. But again, I would argue, to put that simple associative mechanism to use requires specially designed systems that can channel the relevant information into it, and at appropriate times. Even mirror neurons are smart in this respect. They don't resonate willy-nilly with any old action; they only resonate with actions that have a purpose, defined not by the movements themselves but by the movements in a particular context, by some deeper understanding.

Human chameleons

I can hear a chorus of skeptics rising up even as I write this. "Imitation in humans isn't always smart!" they are saying. It's true, at least to some extent. We know that humans tend to imitate speech patterns, uptalk, for example—although we clearly recognize on some level that doing so is not critical for communication (unlike the reproduction of, say, phoneme sequences). Meltzoff's experiment with newborns also seems to point to a reflexive-like imitation capacity (but see below). And there's more.

You may have noticed that in some social situations people tend to mimic each other's postures and gestures. If one person crosses his or her arms or legs, twiddles with his hair, adjusts her glasses, takes a sip of tea, or whatever, the other tends to follow. Teenagers are notorious for mimicking the behavior patterns of their peers. Older married couples are judged to look more alike after they've been together for decades than when they were newlyweds. A proposed explanation for this phenomenon is that the couples tend to mimic each other's expressions, which in turn chisels similarities into facial features.[40] Woody Allen turned the phenomenon into his 1983 film, *Zelig*, a mockumentary about a fairly nondescript man who takes on the appearance and characteristics of those who surround him—a kind of human chameleon.

Formal experiments have confirmed that humans indeed tend to mimic others unconsciously in some social situations. In one such study,[41] participants were brought into the lab and asked to describe a set of photographs. It was explained that they would work in pairs, taking turns describing what they saw in the photographs and that their partners would swap in and out. In reality, the pictures had nothing to do with the actual experiment and the study's "other participants" were confederates, part of the experimental team. After the cover story was fed to the subject, a confederate entered the room and the pair (the subject and the confederate) commenced describing their set of three pictures in turn, and then the first confederate left and a second confederate entered for another round of picture description.

The real experiment involved the nonverbal gestures that the confederates modeled during the turn taking, which was either face rubbing or foot shaking. The first confederate modeled one behavior and the second modeled the other so that each subject was exposed to both gestures. The question was, would the participants rub their faces more when they were working with the face rubber and shake their foot more when in the presence of the foot shaker? The answer was yes, and they did so unconsciously according to post-test debriefing sessions.

In a follow-up experiment, the study team puposefuly imitated the gestures of the subjects during a similar picture description task. After the session, subjects were asked to rate how much they liked their partner and how smoothly the exchange went. Being imitated led to higher likability and smoothness ratings. It's called the *chameleon effect*, the tendency for humans to blend in with their social environment by mimicking, in a largely unconscious manner, the behavioral patterns of those around them.

The chameleon effect seems to suggest that Meltzoff is right about Homo imitans, that we are innately wired to imitate, period. Then with such a foundation we can build complex cognitive abilities. But we run into the same problems with chameleon imitation that we ran into with other types of imitation. Human chameleons don't imitate everyone all the time.[42] While they may tag along in a face-rubbing or foot-wagging fest, they don't shadow their partner's speech and they don't reach for the same pen at the same time. If humans did imitate everyone all the time, it would be creepy, not useful. In fact, excessive and inappropriate imitation, known as *echophenomena*, is a recognizable class of neurological disorders. *Echolalia* is the term for excessive imitation of speech and *echopraxia* is the term for excessive imitation of actions.[43]

Paradoxically, unconscious mimicry is, in fact, highly selective.[44] Humans tend to imitate people they like and distinctly avoid imitating people they don't like. And imitation is more likely among "ingroup" members (such as within nationality, religious, or ethnic group) than among "outgroup" members. Again, having the raw capacity to imitate isn't enough. *Something more is needed* to explain why some gestures are mimicked and others are not.

To understand the chameleon effect we need to consider what functions the behavior might serve. What is the underlying mechanism that guides the imitative behavior? We don't have to look far. Woody Allen seems to have hit on the answer a few years before the social psychologists propelled it to mainstream psychological theory. In Allen's screenplay, Zelig's chameleon behavior was motivated by a deep longing for social approval, which he attempted to gain by transforming himself in the image of those around him. It is a common enough phenomenon (on a less dramatic scale), especially in teenagers but also quite evident in adults if you pay attention. Psychologists Tanya Chartrand and John Bargh put it this way:

> We suspect that the chameleon effect contributes to effective behavior coordination among members of a group. . . . the positive effects of empathy, liking, and bonding that occur automatically because of the chameleon effect would likely benefit most newly formed groups in which relationships among the members do not yet exist or are fragile—it would also tend to shape initial feelings among group members in a positive direction.[45]

In short, unconscious mimicry serves a social function. People who act like me are perceived as part of my ingroup and this breeds positive social attitudes. Or on the flip side, if I have positive social attitudes toward an individual or group, this triggers tendencies to mimic them. Some bit of social machinery is at the core of the process, not the mimicry itself. It's not that we are social because we mimic; we mimic because we are social. This is not just an alternative framing of the relation. The two perspectives have different explanatory power. If we try to use unconscious mimicry as the innate foundation on which sociability naturally grows, we have no explanation for why we don't mimic everyone all the time. How can unconscious, "automatic" mimicry be so selective? If, however, unconscious mimicry is a consequence of a social brain, then social states (whatever they might be—I'll leave that question to the social psychologists) can serve as the "something more" that guides imitation.

BORN TO IMITATE?

IT'S WORTH spending one paragraph on a final objection to the view promoted here that human imitation is prodigious not because we are uniquely built to imitate but because we have the cognitive systems to take full advantage of imitation. Newborn infants seem to have an inborn proclivity to mimic. Maybe there is a developmental argument for how imitation becomes selective: it starts out general and becomes shaped by experience. The demonstration of early imitation in humans anchors this perspective. But the finding hasn't gone unchallenged. Psychologist Susan Jones, for example, has taken a critical look at the data as well as the methods used to gather them, including the jaw-dropping demonstrations of maternity ward mimicry, and came to a rather different conclusion:

> The evidence is consistent with a dynamic systems account in which the ability to imitate is not an inherited, specialized module, but is instead the emergent product of a system of social, cognitive and motor components, each with its own developmental history.[46]

The foregoing discussion was just a long-winded way of saying what Jones said so elegantly and succinctly.

The *mechanism* for imitation is clearly primitive and "embrained," dare I say, in macaque mirror neurons. As Heyes argues convincingly, it isn't a complicated mechanism at its base. All it takes is association. The trick is knowing what and when to associate. For this we need to pair Heyes's associative mechanism with Jones's higher-order systems that can make use of imitation "intelligently" (in the evolutionary sense). Macaque brains are set up with the machinery to take advantage of information about others' actions to inform and select sometimes similar, sometimes different actions of their own. Mirror neurons are part of this machinery. Macaque brains, however, are not set up to take full advantage of imitation for language learning or social networking. Human brains, on the other hand, are built to take better advantage of what imitation can offer. We have the same fundamental associative mechanism, we probably have the same type of mirror neurons, but we differ in the cognitive (information processing) mechanisms that have evolved to put those fundamental associative mechanisms to good use. And that is why humans ape better than apes ape.

9

Broken Mirrors

*To be an autistic child means, with variable degrees of severity,
to be incapable to establish meaningful social communications
and bonds, to establish visual contact with the world of others,
to share attention with the others, to be incapable to imitate
others' behavior or to understand others' intentions, emotions,
and sensations.*

—VITTORIO GALLESE, 2006

AUTISM SPECTRUM disorders are complex and highly variable with a poorly understood cause. There is a very large literature and much debate on possible genetic and environmental causes and an equally large literature attempting to sharpen the diagnosis, identify diagnostic markers, differentiate subtypes of the spectrum, and characterize the source of the variability among affected individuals. Many questions remain unanswered. What causes autism? Why is the incidence increasing? Is it one or many disorders? Why are males more likely to be affected? Are there effective treatments? One could write an entire book on the ins and outs of autism and indeed several such books already exist.

Here my focus is more circumscribed. I restrict the discussion to the behavioral symptoms of autism and (neuro)cognitive models for explaining those symptoms. I highlight two of the most influential hypotheses, the broken mirror theory and the broken mentalizing theory (or broken theory of mind theory—I use the terms interchangeably). Further, I have no intention of providing a thorough review of the host of experiments that have investigated the range of abilities and disabilities in autism or even provide much depth in my discussion of the cognitive theories themselves. Please consult any of the many primary sources for a broader view.[1]

Instead I have two main goals with this chapter. One is to address the basic mirror neuron–based account of autism because the theory has been rather influential and a lot is at stake given how many lives autism touches. The other goal is to highlight an alternative perspective on autism in the same way that (I hope) I've been able to highlight alternative perspectives on mirror neuron function, embodied cognition, and imitation. Specifically, I'm going to suggest the possibility that the dominant neurocognitive theories of autism, which assume that behavioral deficits result from *lack of* or *diminished* social sensitivity, have it wrong and in fact have it backward.

WHAT CAN WE INFER FROM BEHAVIOR?

"DEFICIT THEORIES" of dysfunction are reasonable and intuitive. If an individual fails to respond normally to sound, it's a good bet that the person has a *diminished* capacity to process and hear sound. He simply isn't capable of perceiving the signal. Likewise, if another individual fails to respond normally to social stimulation, it's a reasonable bet that the person has a *diminished* capacity to process social information. But consider the following thought experiment. Imagine you had a stadium rock concert–type sound system hooked up to your living room television and you attempted to watch the evening news with the sound cranked up all the way. Most likely, you would cover your ears and quickly leave. If you forced yourself to stay, you would run into at least one of three problems as you tried to listen and watch. One, the physical pain would be so extreme that you wouldn't be able to concentrate on the message. Two, attempts to dampen the sound and ease the pain, say by sticking your fingers in your ears, would filter out many of the fine details you need to hear normally. You would perceive less well. Three, if you did manage to listen, the extreme volume would excite so many nerve fibers that it would drown out the details of the signal itself and again you would miss many things. Excess can be as detrimental to normal function as paucity.

Consider also the behavioral patterns of two hypothetical individuals. Owen frequents the park or the beach, hiking, biking, swimming, or sailing. He's an outdoor kind of guy with the tan and sun-bleached blond hair to prove it. Ethan, on the other hand, is usually found indoors. He spends his spare time reading, composing, working out at the gym, or enjoying the occasional indoor paintball battle. Ethan's closet contains a rack of long-sleeve shirts and a clutch of wide-brimmed hats hanging on hooks right next to his basket of SPF 70 sunblocks, just in case he does find himself outdoors. Of the two, which is the sun lover? Owen, right? Not necessarily. Ethan could very well crave the warmth and brilliance of the sun, but can't satisfy his desire because he is excessively sensitive to it, while Owen may be indifferent to whether it is bright and sunny or cool and overcast; as long as he's on his bike or boat he's happy.

We can play the same game with food-related behavior. Owen's diet does not include any dairy products, whereas Ethan's staples are pizza and milkshakes. Is it safe to assume that Ethan likes dairy more than Owen? No. This *could* be true, but it is also quite possible that Owen loves cheese and ice cream, more so than Ethan in fact, but is lactose intolerant.

Behavior does not *automatically* reveal its cause and can be misleading. We must keep this in mind as we consider the behavioral phenotype associated with autism and the standard interpretations of it.

AUTISM: THE CLINICAL PERSPECTIVE

THE FIFTH edition of the *Diagnostic and Statistical Manual of Mental Disorders* (DSM-V) characterizes Autism Spectrum Disorder (ASD) by two classes of behavior:[2]

A. *Persistent deficits in social communication and social interaction across multiple contexts,* including (1) deficits in social-emotional reciprocity such as reduced sharing of interests or emotions and failure to initiate or respond to social interactions; (2) deficits in nonverbal communicative behaviors used for social interaction, such as abnormalities in eye contact and body language or deficits in understanding and use of gestures, and lack of facial expressions and nonverbal communication; (3) deficits in developing, maintaining, and understanding relationships, such as difficulties adjusting behavior to suit various social contexts, making friends, and absence of interest in peers.

B. *Restricted, repetitive patterns of behavior, interests, or activities,* including (1) stereotypes or repetitive motor movements, use of objects, or speech; (2) insistence on sameness, inflexible adherence to routines, or ritualized patterns of verbal or nonverbal behavior; (3) highly restricted, fixated interests that are abnormal in intensity or focus; (4) hyper- or hyporeactivity to sensory input or unusual interest in sensory aspects of the environment.

The DSM-V further allows for degrees of severity in the behavioral symptoms and allows for variants with and without accompanying intellectual impairment, and with and without accompanying language impairment, thus defining the *spectrum*.

For our purposes here, we note that the diagnostic criteria for ASD all but assume a *deficit* theory—"Persistent *deficits* in social communication"—and underscore the possibility of "*hypo*reactivity to sensory input." There is a widespread view that autistic individuals are less sensitive to pain because their reactions to painful events, such as a blood draw, are often less dramatic than a typical child's. A recent study[3] took a closer look at autistic individuals' pain reactivity and came to a rather different conclusion. Seventy-three children and adolescents with autism were studied along with 115 controls while they underwent a blood draw. Behavioral pain reactivity to the needle prick was rated by a nurse and a child psychiatrist, who were present during the procedure. Overall the autistic group showed a reduced behavioral response to pain, consistent with popular belief. However, the team also measured the children's heart rate and blood serum levels of a stress hormone (β-endorphin). Autistic subjects showed a greater heart-rate response to the needle prick and had a higher concentration of β-endorphin than controls. The authors conclude, "The results suggest strongly that prior reports of reduced pain sensitivity in autism are related to a different mode of pain *expression* rather than to an insensitivity" (emphasis mine). Again: Behavior does not automatically reveal its cause.

AUTISM: THE COGNITIVE PERSPECTIVE

IN TERMS of cognitively oriented research, autistic individuals appear to lack the ability to, or at least have difficulty with, understanding the goals and intentions behind others actions, reading others' emotions, empathizing, sharing attention with others (pointing out objects of interest or orienting to what others are attending to), recognizing faces, and imitating others. Certainly, there is ample evidence that autistic individuals can perform abnormally on tests of such abilities.

Many of these deficits include the kinds of things that a theory of mind should enable, like understanding the intentions of others, empathizing, and sharing attention. Accordingly, one prominent cognitive account of autism is that affected individuals lack a theory of mind, the ability to "mentalize."[4] Psychologist Simon Baron-Cohen has termed the condition *mindblindness*.[5] It's a reasonable theory: if someone performs poorly on tests that are designed to assess, say, intention understanding, then it is a good bet that the mental module underlying such an ability is broken.

Interestingly, mirror neuron codiscoverer Vittorio Gallese considers this theory "totally untenable," citing as a key bit of evidence a single case study "carried out on a patient who suffered a focal bilateral lesion of the anterior cingulate cortex (ACC), previously identified as the candidate site for the Theory of Mind Module, [who] showed no evidence of mind reading deficits."[6] Gallese proposes his own theory:

> My hypothesis is that these deficits, like those observed in the related Asperger Syndrome, are to be ascribed to a deficit or malfunctioning of "intentional attunement" because of a malfunctioning of embodied simulation mechanisms, in turn produced by a dysfunction of the mirror neuron systems.[7]

Intentional attunement is a fancy term for the ability to tune into the intentions of others, effectively a new label for the mentalizing idea. But Gallese goes further and claims that deficits in mentalizing are not a result of damage to a special-purpose theory of mind neural module, but to a more basic simulation mechanism that is dependent on mirror neurons. This is the *broken mirror theory of autism* that has been put forward by several authors.[8]

We should apply Gallese's lesion-based argument and note that hundreds of patients with Broca's aphasia and/or limb apraxia, who often have lesions that disrupt motor ability and destroy the mirror system, are not reported as being autistic. By Gallese's own logic, one would think that such a fact would render his theory "totally untenable," but the broken mirror hypothesis has proven highly influential nonetheless.

Part of the appeal of the broken mirror theory is that it accounts for high-level social deficits in terms of a basic neural mechanism that covers much cognitive ground. Mirror neurons support language, imitation, theory of mind, empathy, intention or goal understanding, and the ability to read others' emotions. All of these are impaired in autism, therefore autism can be traced back to a deficit in the mirror neuron system. Of course, if mirror neurons don't support any of these functions, as I've argued at length in this book, then the inferential chain that leads to the broken mirror hypothesis falls apart. Based on the arguments put forward in earlier chapters, it follows that the broken mirror theory of autism is itself totally untenable.

I arrive at this conclusion by unraveling the foundational assumptions underlying mirror neuron function. Others have attacked the broken mirror theory from another direction. Cognitive neuroscientist Antonia Hamilton has carefully reviewed the predictions of the theory against available experimental data regarding autistic individuals' behavior. She comes to the same conclusion: autism is not caused by broken mirrors.[9] Hamilton's argument boils down to the fact that autistic individuals simply don't exhibit the behavioral profiles that the broken mirror theory predicts. For example, while it is true that some studies found that autistic individuals perform more poorly on tests of the imitation of *meaningless* actions, other studies found that they can imitate goal- or object-oriented actions quite effectively. And there is also good evidence for preserved action recognition ability in autism. In one study, Hamilton and her colleagues asked autistic children to match pictures of different hand postures to drawings of actions (such as ironing a shirt) in which the hands of the actor were not depicted. The autistic group outperformed the control group of nonautistic children.[10] These abilities should be impaired according to the broken mirror theory.

Psychologist Morton Ann Gernsbacher also reviewed the relevant empirical evidence and reinforced Hamilton's assessment. She writes that a number of studies "are unanimous in demonstrating that autistic individuals of all ages are perfectly able to understand the intentionality of their own actions and of other humans' actions; there is neither 'incapacity' nor impairment in understanding of the intentions of action."[11] Gernsbacher also raises serious concerns regarding the neuroimaging data argued to reveal dysfunction of the mirror system in autism. She points to two studies that reported abnormal mirror neuron activity in autistic subjects in 2005 and 2006.[12] The studies were highly publicized in outlets like *The New York Times*, *Scientific American*, and *NOVA*, yet, Gernsbacher notes, their results failed to replicate (could not be reproduced in subsequent studies) and some subsequent studies reported no difference between autistic and control groups.[13]

Psychologist Cecelia Heyes also argues against the broken mirror hypothesis. Her argument centers on imitation, which according to the broken mirror claim should be severely impaired in autism. Heyes rejects the commonly accepted notion that autistic individuals have a fundamental deficit in imitation. Her argument is that standard tests of imitation are not direct assessments of imitation ability because of their cognitive complexity. In typical tasks, participants are asked to "do this" while some action is modeled, which requires conscious inferences about what specifically is being asked, sustained attention, working memory to remember the sequence of movements, and the motivation to comply. To get around these issues, Heyes and colleagues measured automatic imitation (unconscious mimicry) by asking autistic participants to open or close their hand as quickly as possible when they saw a stimulus hand start to move. The team wanted to know whether autistic individuals would show a movement compatibility effect: faster response times when the perceived action matched the required response. This is precisely what they found, indicating that autistic individuals are quite capable of associating perceived and executed actions.[14]

Overall, and despite the hoopla surrounding the broken mirror theory of autism in the popular press, there is a growing consensus that the scientific evidence does not support the claim.

What about the broken mentalizing theory? The strongest argument for a theory of mind deficit in autism comes from the *false-belief task*, a classic behavioral assay for the ability to mentalize. In the standard version, one puppet, call her Sally, places a desirable item such as a piece of chocolate in a basket and then leaves the scene. Another puppet, Anne, then enters and moves the object to a nearby box. Sally then returns and the subject, who had been watching the drama, is asked to predict where Sally will look for her object. The key bit is that in order to get it right, the observer has to recognize that Sally *believes* that the object is still in the basket even though it is not. If the subject predicts that Sally will look in the basket, this means that he or she knows that behavior is caused not by the environment alone (where the object is, in fact) but by people's *beliefs* about the world, that is, mental states. The false-belief task is argued to be, therefore, a litmus test for the presence of theory of mind.

Four-year-old children pass the false-belief test but children younger than four usually fail, as do autistic children and autistic adults (more often than controls, anyway). From this it is typically concluded that autistic people and nonautistic children under the age of four lack a theory of mind.

Paradoxically, these facts immediately raise problems for the idea that the behavioral symptoms of autism are caused *fundamentally* by a broken theory of mind module. For one, as Yale psychologist Paul Bloom points out, three-year-old children, who cannot pass the false-belief test, do not behave as if they are autistic.[15] In fact, autism is often diagnosed well before the child's fourth birthday, an age at which autistic and nonautistic children are indistinguishable in terms of performance on such tests of theory of mind. Another problem is that a nontrivial proportion of autistics and many higher-functioning individuals on the autism spectrum *can* pass the false-belief task, but still exhibit the social difficulties that define the spectrum disorder.[16] There must be something beyond a theory of mind deficit, as measured by the false-belief task, at the core of autism.

The false-belief task itself may be part of the problem. Gernsbacher noted that success on the task depends in large part on language ability: to get the answer right, you have to understand the question you are being asked. Given that autistic individuals have language problems, it is not surprising that they fail the false-belief task.[17] Bloom has similar qualms about the task and even articulated two reasons for abandoning it altogether as a measure of theory of mind.[18] His first reason is that the task is too cognitively demanding and therefore isn't tapping into just theory of mind. Similar to Heyes's argument regarding imitation tasks, Bloom notes that the false-belief task places significant demands on the ability to suppress salient but misleading information for correct task performance (like where the object actually resides), and so on. In support of his claim, he points to evidence that with simplified versions of the test, kids younger than four can get it right. His second reason is that theory of mind is required for other tasks that children even younger than two years old typically pass. We already encountered one example in the previous chapter: toddlers often imitate the *intended* rather than the actual action: if an experimenter attempts to operate a toy but fails, children imitate the *intent,* not the failures, demonstrating that very young children understand intentions. Bloom cites several more examples.

Though not extensive, there is some evidence that, as with children younger than four years old, simplifying the theory of mind task reveals the capacity to mentalize in autism. One task involves a doll that favors a snack that the participant dislikes. The child is then asked to predict which snack the doll will choose to eat. Autistic children who fail the standard false-belief task can often pass the simplified task.[19]

So neither the broken mirror nor the broken mentalizing theory hold up well to the empirical facts regarding the abilities and deficits in autism. Antonia Hamilton concludes her review of the literature:

Neither a low-level theory (the broken mirror theory) nor a high-level theory (the broad mentalising theory) can fully account for the current data. It is likely that future theories will need to be more subtle and to distinguish between different types of mirror neuron system and different types of mentalising. Thus, blanket statements about deficits in "the mirror neuron system" or "mentalising" in autism will no longer be sufficient.[20]

I agree that neither theory is satisfactory, but I'm not convinced that more subtle distinctions between types of mirror system or theory of mind operations will fare better. The problem, I suspect, is hidden in the fact that all of this discussion still centers on ideas about what is *lacking* in autism. Autistic people have no mirror system or no theory of mind or no empathy or no ability to process social information. These are *deficiency* or *hypo*function theories; a good first guess, but not the only possibility. And given that they haven't had all that much success, maybe it's time to focus some research effort on a theory based on *excess* or *hyper*sensitivity. Perhaps autistics don't experience a socially numbed world but rather a socially intense world.

THE INTENSE WORLD SYNDROME

IN FACT, Henry Markram, Tania Rinaldi, and Kamila Markram proposed such a theory in a 2007 article aptly titled, "The Intense World Syndrome—An Alternative Hypothesis for Autism."[21] The theory is grounded, oddly enough, in a rat model of autism. I say "oddly enough" because autism has traditionally been considered a uniquely human disorder, with defining symptoms showing up in high-level social and language domains. But it turns out that rats who are exposed prenatally to valproic acid—a compound used in human medications to control seizures and bipolar disorders—develop some key features of autism both neurally and behaviorally, including loss of cerebellar neurons, abnormalities in the serotonergic system, decreased social interactions, increased repetitive behaviors, enhanced anxiety, motor abnormalities, and sensory hypersensitivity. Curiously, the prevalence of autism in humans who are prenatally exposed to valproic acid through maternal use of the medication is substantially higher (one estimate is 11–100 times) than in the general population.[22]

Some of the symptoms listed above—increased anxiety, sensory hypersensitivity—don't regularly surface in the debates over broken this or broken that theories of autism, and for good reason: the theories have little to say about them. How do you end up with hypersensitivity to sensory stimulation in a syndrome caused by a broken theory of mind module? But these symptoms are fairly typical of autistic individuals and something that we should seek to explain. As we'll see, hyperfunction accounts have a natural explanation.

The existence of a rat model makes it possible to explore, in substantial detail, the neural bases of an autistic-like phenotype. The Markrams and their team did precisely that. They focused on three brain regions, the somatosensory cortex, the prefrontal cortex, and the amygdala. Somatosensory cortex is well developed and extensively studied in rodents and so provides a good model for sensory systems. They assessed prefrontal cortex as a means to sample a higher-order cortical processing system. And they studied the amygdala due to its known role in emotional processes and to assess neural function in a nonneocortical network. Here's the highlight reel version of what their work showed.

They determined that local neuronal networks in the three brain regions tested in rats are *hyperreactive*. They demonstrated this by broadly stimulating, in vitro, a small patch of rat brain tissue (less than one millimeter square) and recording the response from individual neurons embedded within that patch. The broad stimulation activates both the recorded cell itself as well as its local network. The response, therefore, reflects not only direct stimulation but also indirect network stimulation via the cell's neighbors. In animals treated with valproic acid, the neuronal response to network stimulation was greater than in control animals. In fact, it was about twice as strong in the two cortical regions. Interestingly, this is not due to the intrinsic hyperreactivity of the cells themselves. If a single cell is stimulated rather than its network, *hypo*responsivity occurs: more stimulation is required to make the cell fire. So it is not the neurons themselves that are hyperexcitable, it's the *networks*. The driving force behind the micro network–level hyperreactivity appears to be the direct connections between neurons. Valproic acid–treated animals exhibit a 50 percent increase in the number of direct connections between neurons within a local circuit.

Further, the Markrams and their team found that neural networks in valproic acid–treated rats are also *hyperplastic*. It has been known since the 1960s that if two connected neurons are stimulated simultaneously for a period of time, the subsequent response to stimulation of the two-cell network is enhanced for quite a long time afterward compared to baseline. This is *long-term potentiation* (or LTP) and it is thought to be an important cellular mechanism in learning and memory. The LTP effect doubled in valproic acid–treated animals compared to control rats in all three brain regions studied, showing that this form of plasticity is enhanced. Network connectivity plasticity is also enhanced. When connectivity patterns between neurons in local and expanded networks were examined before and after widespread and prolonged (overnight) activation of the entire network, valproic acid–treated rats exhibited an increase in the rate of rewiring, mostly evident in the nonlocal networks, compared to controls. This hyperplasticity seems to have behavioral consequences: treated rats have an exaggerated fear learning response—response to a tone that had previously been paired with a mild electric shock—that is stronger, generalizes more readily, and persists longer than in control animals.

Given these kinds of neural changes in the rat model of autism, a plausible story can be told about the neural basis of the range of autistic behaviors. Hyperabilities, such as increased sensory sensitivity or memory in specialized domains, can be explained by hyperreactivity and hyperplasticity of neural circuits. Hyporesponse to social stimuli can be explained in terms of the emotional intensity of the signal, which triggers anxiety and avoidance responses, which means less information is acquired in individual social situations and over time reduced opportunities to learn in the social domain. Theory of mind performance would also be expected to suffer with a hyperactive response to social signals—even if there is no fundamental deficit in mentalizing—because of increased anxiety when interacting with others and/or because avoidance behavior decreases the amount of information perceived or learned. Repetitive behaviors can be viewed as a coping mechanism aimed at regulating the child's intense world. Motor deficits can be explained by hyperexcitability of the response to sensory stimulation, which has motor consequences, as we've discussed, or from hyperreactivity of motor systems themselves. And because language is at some levels a sensorimotor task and at other levels a highly social behavior, abnormalities in the sensorimotor or social domains can be expected to affect language.

All of this is interesting—suggestive even—but how relevant is it, really, to autism? That's a fair question. At this stage, all we really have is circumstantial evidence in the form of parallels between human autism and the valproic acid–treated rat model. Ultimately, the proof of the pudding is in the eating. If therapies or preventative measures can be developed in the rat model that actually prove useful in humans, or if the same neural changes could be unambiguously demonstrated in humans, then we have a decent pudding.

But even if we don't have a perfect pudding in the end, the rat model of autism is nonetheless useful for our purposes here because it essentially demonstrates proof that autism-like symptoms can emerge from a hyperreactive system. It doesn't have to be the case that abnormal performance on this or that task is a result of depressed neural function. The Markrams make this point eloquently in connection with previous work on the link between amygdala dysfunction and autism. It is worth an extended quote (citations removed):

> The very first animal model of autism was based on lesioning the amygdala and studying the effects on social behavior and hierarchy, implying that the lack of amygdala activity may explain the lack of social interactions or social intelligence in autism. This view dominated the research performed on the role of the amygdala in autism. Parallels were drawn between amygdala lesioned patients and autistic subjects, functional magnet[ic] resonance imaging (fMRI) studies revealing an insufficiently activating amygdala in autistic subjects were associated with deficits in interpreting other people's state of minds and feelings. However, the opposite could also be true and lead to similar symptoms: rather than being hypo-active or not sufficiently responding, the amygdala could be overly reactive in

autism. Consequently, autistic people could be processing too much emotionally relevant information, including enhanced fear and anxiety processing. The outcome could be a similar one to a not sufficiently active amygdala: withdrawal and decreased social interaction due to an enhanced stress-response and socio-emotional overflow. Indeed, as described below our studies on [valproic acid]-treated rat offspring indicate that the amygdala is hyper-reactive, hyper-plastic, and generates enhanced anxiety and fear processing. In accordance with this, more recent fMRI studies as well reveal amygdaloid hyper-activation in autism.[23]

This kind of effect—hyper-responsivity leading to avoidance—is observed regularly and uncontroversially in the sensory domain.[24] Autistic individuals often cover their ears when even moderately loud sounds are present in the environment and exhibit other forms of avoidance behavior. As with the rock concert sound system example at the beginning of this chapter, if an autistic person failed to get information out of moderately loud sounds or simply left the room, we wouldn't say that he or she had a diminished capacity to hear the sound. The response is more readily explained as an increased sensitivity to sensory stimulation. As autistic author Temple Grandin said in a radio interview, "How is a person going to socialize if their ears are so sensitive that just being at a restaurant is like being inside the speaker at a rock 'n' roll concert and it's hurting their ears?"[25] Good question.

DO YOU SEE WHAT I SEE?

ONE OTHER indicator of hypersensitivity is staring us in the face, literally. Autistic individuals seem to perceive less in facial expressions than nonautistic individuals, and the part of the brain that is partial to faces, the fusiform face area, responds less well in autistic folks. The first-pass (and most popular) interpretation of these findings: autistic people can't read faces because their neural face area is poorly developed. A research team in Pittsburgh scanned Temple Grandin while she was watching pictures of faces and nonface scenes and found exactly this result. Grandin described her experience in the scanner while lecturing at UC Davis's MIND institute in 2008 (CAPS indicate emphasis in Grandin's phrasing):

> Now [researcher] Nancy Minshew did another brain scan and she found I was more interested in THINGS than I was in looking at pictures of people. . . . She starts showing me all these weird videos of people, airplanes flying over the Grand Canyon, bridges and apples and all kinds of objects. And I'm looking at this [thinking] *where did she get this 1970s video? How many copyright violations do we have on this video?* Why was I looking at the THINGS? Because the THINGS told me more information about where the tapes came from. And I was trying to figure out what the experiment was all about.

If all you did was analyze the brain scans it would look like Grandin's face area is dysfunctional. But, it is clear from her recounting of the experience that she was attending more to the nonface pictures, which could easily produce the observed results even if her fusiform face area (FFA) were perfectly normal (attention modulates the neural response). You are wondering whether the reason why she was more attentive to the nonface pictures is because her face area is dysfunctional in the first place. It's possible, and it is the standard explanation of both the brain response pattern and Grandin's behavior, which is more object- than face-oriented. It's not the only possibility, though! It could be, for example, that her FFA is *hyper*reactive, which leads her to avoid attending to faces, which results in more attention paid to nonface objects, which leads to the observed imaging result. Or maybe she's just smart enough to recognize that the THINGS tell you more information, as she noted, in the context of the problem she had tasked herself with during the study, to figure out the goals of the experiment.

At least one study has confirmed that alternative explanations of the face processing "dysfunctions" in autism may be on the right track. Autistic and nonautistic individuals were scanned using fMRI while they looked at pictures of faces that were either emotionally neutral or emotionally charged. Crucially, using eye-tracking technology, the researchers also monitored which parts of the images their participants were looking at during the experiment. Overall, autistic participants activated their fusiform face region less vigorously than nonautistic controls, replicating previous work. But the eye-tracking data showed that this was simply because they spent less time looking at the most informative region of the faces, the eyes. In fact, when the researchers looked at fusiform activation as a function of time spent fixating on the eyes in the photos, they found a strong positive correlation in the autistic group. This means that the autistic brain is responding quite well to face stimuli, if one takes into account the amount of time spent looking at them.[26]

Again we might ask the same question we asked previously, why aren't autistic individuals looking at the most informative region of a face in the first place? If it's not a general face processing deficit, maybe it is a facial *emotion* processing deficit that limits their ability to detect information in the eyes. According to this view, the face processing system *in general* is working OK, reflected by the activation in the FFA when autistics actually look at faces, but because autistic people can't process the emotion in them, they don't spend as much time looking at the critical regions compared to controls. As before, this is a possible interpretation. But again, it's not the only interpretation. An alternative is that autistics don't look at the eyes as much because of a hyperactive response to emotional information, which is particularly evident in the eyes. And consistent with this alternative possibility, the same study reported that amygdala activation was stronger in the autistic compared to the nonautistic group while looking at faces.

Also consistent with the alternative, emotional hyperreactivity hypothesis are statements from autistic individuals themselves. Here's a sample gleaned from a paper covering face processing in autism:[27]

It's painful for me to look at other people's faces. Other people's eyes and mouths are especially hard for me to look at. My lack of eye contact sometimes makes people, especially my teachers and professors, think that I'm not paying attention to them.

—MATTHEW WARD, STUDENT,
UNIVERSITY OF WISCONSIN

Eyes are very intense and show emotions. It can feel creepy to be searched with the eyes. Some autistic people don't even look at the eyes of actors or news reporters on television.

—JASMINE LEE O'NEILL, AUTHOR

For all my life, my brothers and everyone up 'til very recently, have been trying to make me look at them straight in the face. And that is about the hardest thing that I, as an autistic person, can do, because it's like hypnosis. And you're looking at each other square in the eye, and it's very draining.

—LARS PERNER, PROFESSOR,
SAN DIEGO STATE UNIVERSITY

These are revealing statements for two reasons. First, they provide a clear indication of an intact theory of mind in these individuals ("my lack of eye contact . . . makes people . . . *think* that . . ."). And second, active avoidance of eye contact provides just as much evidence for sensitivity to the information contained therein as does active engagement of eye contact. If you can't recognize that there is information in the eyes, why avoid them?

———•———

AUTISM REMAINS poorly understood on all levels. I'm not trying to suggest that I have the answers. We do know a few things though. We know that the broken mirror hypothesis does not fare well in light of empirical facts. Autistic individuals do not exhibit the range of deficits that the theory predicts and more importantly in my mind, mirror neurons do not support the kind of abilities that are affected in autism. The broken mentalizing theory doesn't do much better given that there are significant concerns about the tasks that are typically used to assess theory of mind, the fact that performance on such tasks does not reliably distinguish autistic from nonautistic individuals, and the fact that a broken mentalizing account only addresses one aspect of the overall picture in autism. The intense world theory shows some promise. The range of data we've considered in this chapter more than justifies ramping up the investigation of this theory in the "human model" of autism.

———◆———

AN ASIDE: Just because someone doesn't make eye contact, slinks away from social interaction, or appears overly obsessed with THINGS, doesn't necessarily mean that he or she can't read emotions, is *a*social, or lacks empathy. Conversely, just because another person is engaging, gregarious, and charismatic doesn't mean he or she has achieved the platonic social ideal. Sociopaths, for example, sometimes exhibit just these features; serial murderer Ted Bundy is a famous example.

In the wake of the recent rash of shootings in Aurora, Newtown, and elsewhere there has been an unfortunate implied or explicit link made between autism and antisocial behavior. The link is easy to make under standard assumptions about the disorder, particularly high-functioning autistic individuals who are described as being not like us, an enigma, or lacking in empathy. They've been deemed robotic and alien, even likened to nonhuman animals, rendering the ingredients for the classic Hollywood demon. Experts repeatedly and correctly point out, however, that there is no causal link between autism and violence, nor between autism and sociopathic behavior.

IO

Predicting the Future of Mirror Neurons

RAZING THE BARN

THE MIRROR neuron theory of action understanding—with its forays into language, empathy, theory of mind, autism, and more—has had a good run. In the 1990s, theorists climbed an inferential ladder from the response properties of mirror neurons to some of the loftiest problems in human cognition. Since that time, the vast majority of research on mirror neurons has focused—and continues to focus—not on macaque mirror neurons themselves, but on the applications to humans of mirror neuron theory. Much of this work still assumes that macaque mirror neurons support action understanding.

What many researchers fail to notice, however, is that the inferential ladder has quite probably been kicked away. Human research shows that the ability to understand actions, whether speech, manual gestures, facial expressions, or the movements of snakes or birds, does not depend on the ability to perform those actions with one's own motor system. A simpler explanation of mirror neuron behavior, sensorimotor association for the purpose of action selection, was uncovered (or, if you like, rediscovered). Further, a simple simulation mechanism fails to adequately explain complex human behaviors like language, understanding, theory of mind, imitation, or autism.

Why did the theory become so popular? It had three major appeals. First and foremost was simplicity. You could explain it in one line: "*we understand action because the motor representation of that action is activated in our brain.*"[1] Second, it offered the promise of bringing this simplicity of explanation to many complex problems. And third, it was grounded in hardcore animal neurophysiology, providing not only a neural mechanism for human cognition but also revealing an evolutionary pathway to its development. It felt like mirror neurons had opened a new doorway into understanding the mind, one that offered a simpler explanation than dominant theories at the time.

Yet a great deal of complexity lay hidden in the simple explanation. Mirror neurons don't resonate with every action. How do they "know" which actions to simulate and which to leave alone? Where is *this* understanding coming from? If mirror resonance is the basis of human language, why don't macaques talk? If mirror resonance is the basis of imitation, why don't macaques imitate in the way humans do? If mirror resonance is the basis of empathy, social cognition, and all things human, why don't macaques act more like us? *Something* must have evolved that is causing the human mirror system to behave differently than the macaque mirror system. It's the dark matter of the mirror neuron universe.

This "dark matter" was the topic of study for the last five decades or so in cognitive psychology, linguistics, and related fields, with much progress made. Research coming out of these "traditional" approaches suggests that macaque and human mirror systems behave differently because they are plugged into different computational or information processing neural apps. The human mirror system is wired up to networks that support complex conceptual understanding, language, theory of mind, and the rest. The macaque system is plugged into an impressive information processing system as well, but one that is different from humans. Viewed from this perspective, what we see when we look at the behavior of the mirror system is the *reflection* of the information processing streams that plug into it.

We've been down a similar road before. Behaviorists had very simple mechanisms (association and reinforcement) for explaining complex human behavior. But removing the mind as a mediator between the environment and behavior ultimately didn't have the required explanatory oomph. Mirror neuron resonance theory isn't quite behaviorism, but there are not many degrees of separation because "it stresses . . . the primacy of a direct matching between the observation and execution of action."[2] The notion of "direct matching" removes the sorts of operations that might normally be thought to mediate the relation between observation and action systems, such as categorizing the action, recognizing its behavioral importance, interpreting it in context, and so on. The consequence of such a move is loss of explanatory power. The mirror neuron direct matching claim results in a failure to explain how mirror neurons know when to mirror in the first place.[3] We then have to look to the "cognitive system" for an explanation, which lands us back where we started: with a complex mind behind the mirror neuron curtain of explanation for complex mental functions.

I am sympathetic to the lure of simplicity. As scientists we strive for the simplest theory that explains the most facts. That is our metric of success. Mirror neurons at first appeared to be the theoretical equivalent of a superhero: they used a simple, singular skill to wield extraordinary power over all other cognitive domains and theories. And who doesn't love a superhero? But a closer look reveals that the singular skill of mirror neurons, while important, requires a pantheon of other superheroes to do anything useful.

SEND IN THE CARPENTERS

I'VE FOCUSED nearly exclusively in this book on the theoretical perspective from Parma, and for good reason: the Parma theory is what generated all the excitement. But theories of mirror neuron function are beginning to evolve, indeed have been evolving for the last decade. Authors such as Michael Arbib,[4] Gergely Csibra,[5] Cecelia

Heyes,[6] James Hurford,[7] Pierre Jacob and Marc Jeannerod,[8] James Kilner,[9] and others have put forward alternative perspectives, only a few of which I've discussed in the preceding pages. Some of these viewpoints, like my own, discount the role of mirror neurons in "understanding." Others entertain an intermediate possibility, that mirror neurons can *augment* action understanding; or, to paraphrase a well-known tag line: mirror neurons don't *make* action understanding; they make action understanding *better*.

I'll illustrate using the model proposed by computational neuroscientist Michael Arbib with his collaborator Erhan Oztop. Like Cecelia Heyes's proposal, Arbib and Oztop hypothesize in their "mirror neuron system" (MNS) model that mirror neurons acquire their sensory properties via self-observation.[10] Motor commands are executed via canonical (object-related motor) neurons, which result in limb movements that the monkey itself observes visually. Noting that visual feedback is important for motor control, as the case of Ian Waterman demonstrated (Chapter 7), Arbib and Oztop propose that mirror neurons developed originally for that purpose. Just as canonical neurons use visual object information to select actions, input from self-observed actions can be used, via mirror neurons, to further guide or refine movement control. According to the MNS model, then, mirror neurons don't develop for action understanding, they develop for motor control. *However,* once such a system is in place, Arbib and Oztop argue, it is possible to run the observed actions of *others* through the same system and thus establish the correspondence between self- and others' actions, which can assist in understanding.

How might it assist in understanding? James Kilner and colleagues (among others) argued that it allows the system to make *predictions* regarding future states or goals of the movement because making predictions about the future is a basic property of the way motor systems work.[11] It's a fascinating discovery about how we control our bodies and worth outlining here.

As we saw in Chapter 7, sensory feedback is critical for action. But there's a major problem with such feedback: it's too slow. Imagine you are a subject in an experiment carried out in a motor control research lab. You are seated in a chair, gripping a handle attached to a robotic arm. Your task is to move the handle from one position to another. After a couple of tries you get used to the mild inertia of the device and you can execute the movement quite easily. The experiment then starts. You make the movement over and over until on one trial you feel your arm suddenly swinging wide of its mark. Sensing the deviation, you make a quick adjustment by applying more force in the opposite direction and you are back on track.

We make these kinds of adjustments all the time and with ease. The milk jug is a little heavier than we thought; someone brushes our arm as we reach for our latte; we hit a slick spot on the sidewalk and lose traction. Paradoxically, we are better at compensating for these deviations than we should be.

Consider the task faced by your motor system in the robotic arm situation. At some point in your reach the robotic arm exerted a force of its own, causing your arm to deviate from its course. Your body sense system registers this force, but it takes upward of 100 milliseconds for the signal to reach your brain. Once there, the information has to be analyzed (*how far off are we?*), a correction signal has to be calculated, and then the information has to be sent back down the arm to implement the correction, all of which takes time. Meanwhile, the movement is still ongoing and the arm is in a different position by the time the correction signal arrives at the muscles. In essence, the brain is getting sensory information about what happened to the limb in the *past* (100+ milliseconds ago) and has to generate a correction that will work for the *future* position of the limb when the signal actually arrives. It's like trying to drive a car while looking out the rearview mirror; the only information you get is where you've been, but you have to figure out how to control the car in the immediate future. It's a difficult engineering problem.

Happily, the brain has solved the problem with an ingenious trick. The brain *models* or *predicts* the current and future state of the limb *internally* using the motor commands themselves rather than sensory feedback alone. Motor control scientists call it an *internal model* or *internal forward modeling* ("forward" because it is predicting the state of the limb forward in time); some call it *predictive coding.* Basically, after much experience with moving our bodies and recording the (delayed) sensory feedback from those movements, the brain learns the relation between particular motor programs and how the body responds. Over time, it is possible to predict the outcome ahead of time. Given a system that is capable of making accurate predictions regarding the sensory consequences of movements, actual sensory feedback can be used more efficiently and rapidly because the neural analysis is reduced to error detection: does the actual sensory feedback match the expected (predicted) sensory feedback. Returning to the example of driving a car using only the rearview mirror, imagine how much easier it would be if you had a video "ghost image," a kind of template, of what the rear view *should* be (predicted from past experience) and to stay on the road all you have to do is steer in such a way that the ghost image and the actual image line up. *Predictive coding via internal forward models is the key to efficient motor control and is a major computational function of the dorsal, sensorimotor stream.*

Beside the fact that we are better at movement control than we should be, given the rearview mirror problem, much evidence supports the existence of internal models for motor control.[12] One interesting source is the fact that we can't tickle ourselves. If a tickle feeling is a sensory response, and if the sensory stimulation on your skin is the same when you tickle yourself compared to when someone else does it, why the difference? Predictive coding provides an answer: your brain predicts the sensory consequences of *your* movement, thus decreasing their perceptual impact, but can't predict *others'* movements, so they are more intense perceptually.[13] The same phenomenon occurs whenever we move our eyes. Every time we look left or right, the visual image in front of us sweeps across our retina yet we perceive the world as stationary. If we could reverse this situation—fix your gaze and move the visual image—you would have a clear percept of the world moving. To demonstrate, close one eye and gently push on the side of your open eyeball with your finger. You perceive motion. You are moving your eye, thus causing the visual image to move relative to your retina, which ultimately registers a motion signal. When we move our eyes normally our brain takes that eye-movement motor program and translates it into a prediction for how much change to expect in the visual image; the result is a suppressed sensory perception of the resulting motion, similar to a self-tickle.[14]

There's more evidence. Hearing your own vocalizations *while you are talking* yields a reduced physiological response in auditory cortex compared to hearing a recorded playback of those same vocalizations.[15] When we speak, motor commands for producing speech suppress the response to the expected sensory consequences of those motor acts. Similar effects were reported in another vocal primate, the common marmoset monkey. Single unit recordings in this species' auditory cortex reveal that the firing rate of neurons is suppressed *prior* to self-vocalizations and is selective—some neurons are suppressed for one type of call but less for another.[16]

It seems, then, that sensory prediction is pervasive in motor control and is realized in terms of a *suppression* of neurons coding the predicted event. Why suppression rather than enhancement? It's a computational mechanism for error detection, for noticing when something goes wrong, which is really all the motor system cares about when it comes to sensory feedback. If a reach or a speech act is on target, then all is well with the world from the motor system's perspective; no adjustments are needed. Only when an action misses or is about to miss its mark does the motor system need to pay attention to sensory feedback. The predictive coding neural circuit is set up to implement this information processing goal. It works by inhibiting the activity of the sensory neurons that precisely correspond to the predicted event so that when the event occurs, the normal excitation that would have been caused by sensory stimulation is suppressed— *unless* there is an error in the prediction, in which case the sensory stimulus excites a set of nonpredicted (noninhibited) neurons that register a robust sensory signal. From the perspective of the motor system, the presence of a strong sensory feedback signal means something went wrong, whereas a weak or absent sensory feedback signal means all is well.

You can think of the mechanism as a kind of neural Whac-A-Mole. Picture the moles as an array of sensory neurons turning on (popping up) in response to different patterns of sensory stimulation. It's not hard to imagine that the pattern of mole appearances can be "read off" as a code for the stimulus that is causing them to pop. If we wired up a bunch of moles to a standard QWERTY keyboard, one mole per key, and then someone typed a message on the keyboard, we could learn to read the moles as a code for the letters. Now imagine that *you* are typing on the keyboard trying to communicate a message to someone else who is reading the moles. You want to get your message across accurately so you watch for errors in the mole-popping pattern. But this is complicated. Not only do you have to worry about typing, you have to simultaneously read the moles and evaluate the pattern for errors.

There is a more efficient way to set up the system for error detection, albeit a bit more complicated from an engineering standpoint. Suppose we implant some electrodes in your motor cortex and tap into the motor program indicating which fingers you intend to move, where you intend to place them, and when—the motor code for the letter sequences you *intend* to type. Then we wire up the output of the electrodes directly to the Whac-A-Mole machine such that when the implanted electrodes detect a movement sequence that corresponds, say, to the intended word *mole*, it whacks the moles "coding" m-o-l-e in sequence and *inside* the box before they can emerge: *internal predictive mole whacking.*

Now we have two inputs to the Whac-A-Mole machine, one coming from the keyboard reflecting what you actually type, and one coming from your brain reflecting the motor code for what you intend to type. If the two signals match—if you actually type what the motor program specifies—the moles that would normally pop via keyboard inputs get wacked as a result of your predictive neural inputs. The result is that accurate typing suppresses mole popping.

If you make a typing error, however, the mole wired to the incorrectly typed key pops up, signaling an error. With this setup, error detection is easy: if no moles pop, all is good; but if a mole pops, something is wrong. Notice that you don't need to read the message of the moles to find an error; the errors are automatically highlighted by the mechanism, which is designed to detect deviations. Nor do you have to read the message to make a correction. All you have to do is adjust the movement that went wrong. *This is the computational advantage of predictive coding via neural suppression.*

Now, let's return to mirror neurons. We have a built-in mechanism, internal predictive coding, needed independently for motor control, by which motor signals can influence perceptual networks. This is the mechanism that a number of theorists, such as Arbib and Kilner, would like to tap into as an explanation for how mirror neurons might augment action recognition. In my own theoretical work, and elaborating on the proposals of many others before me,[17] I argued precisely the same point in the speech domain. In a 2011 paper with collaborators John Houde and Feng Rong, I argued that, while mirror neurons can't be *the* basis of speech recognition, they might be able to enhance speech recognition (e.g., identify syllables) by coopting motor-generated predictive coding mechanisms.[18] This, we claimed, might be the neural mechanism behind the small effects on speech perception observed in TMS experiments involving motor cortex stimulation (discussed in Chapter 5).

But I remain a skeptic, even of my own ideas. My concern stems from the computational goal of motor prediction, which is to highlight errors, *deviations from prediction*, via neural suppression. This kind of prediction tends to dampen perceptual sensitivity to predicted events, Whac-A-Mole style, and is not a desirable property for a system that is supposed to *enhance* perceptibility or recognition. In terms of the mirror system simulation mechanism, it means that simulating an observed action in one's own motor system should suppress the sensory signals coming from the event itself and thus decrease its perceptual impact, like a self-tickle, exactly the opposite of the desired effect.

Some would argue this is precisely the point. If the system can make predictions regarding someone else's action via motor simulation, then what is highlighted in the perceptual system is *deviation* from prediction (popping moles), thus simplifying the computational task for perception like the predictive mechanism does for motor control. At first glance, it seems like a brilliant example of neural recycling, taking an efficient computational mechanism designed for one purpose, predictive coding for error detection in motor control, and coopting it for another. A closer look reveals a glitch.

Consider the Whac-A-Mole analogy again. The predictive coding signal that whacks moles prior to their appearance works well to tell us whether our actual keyboard pecking actions match the motor cortex codes that drive the pecks. Does it work for reading the message of others' keyboard pecking? As we watch the machine we see a pattern of moles start to appear, the consequence of someone's typing on a keyboard that is wired up to the Whac-A-Mole machine. The ultimate goal is to read the message coded in the mole popping and we could do this by translating each mole appearance into its corresponding letter (based on prior experience with mole-reading): mole_row_3_column_6 = G, mole_row_2_column_5 = R, mole_row_2_column_4 = E, and so on, thus spelling out words that we can understand, say, *GREEN EGGS*. This would be perceptual stimulus detection (identifying mole pops) and understanding (mole-to-letter/ word translation) *without predictive coding*. But we want to facilitate this process via predictive coding enabled by motor simulation. Here's how it might work.

Since we previously learned the association between our own particular motor codes and the moles they pop, we can run the linkage in reverse and transform an observed mole pop into a motor command by *"direct matching"* (mole action observation makes the corresponding finger twitchy). Now if we execute these reverse-linked motor commands and press our *own* keyboard keys we can *simulate* the key presses of the other person, which generates a mole-whacking *prediction*. And, as we've seen, predictive coding can facilitate processing by reducing the problem to simple error detection. All of this can happen, notice, by direct matching, that is, without having to understand the message.

The problem is immediately apparent. We are only able to link the mole pops into motor commands *after* the moles actually pop. Thus, the motor simulation is doomed to remain one step behind the moles, hardly a *predictive* code, and incapable of facilitating the perceptual task by reducing it to error detection. Unless, that is, the motor sequence that we reverse link to is a familiar one. If it is familiar, then maybe the first part of the sequence (GREEN EGGS) can trigger the second part of the sequence (AND HAM) in advance of the moles. Now when we simulate the key presses of the other person we can get slightly ahead of them, because we know what's coming in the motor sequence, and generate a truly predictive mole whacking code that prewhacks them before they can emerge. This would silence the Whac-A-Mole machine (as long as our prediction is correct) and transform the task from mole detection and letter/ word translation into a simple error detection task, just like it does in motor control.

There is another problem. While we have indeed reduced the task to detecting prediction error, we are predicting and detecting error in *motor plans*—left pinky → right index → left middle—not the content of the *message*, which is what we are trying to understand. Recall that the meaning of an action is not coded in the motor plans. The kicker is that if we are successful in predicting the mole popping sequence via motor simulation, we end up whacking all the moles before they pop and thus squashing the signal. We are left with a correctly simulated motor code (that doesn't contain the meaning because it is directly matched to mole observation) and no moles for the message decoding system to read. Can the message decoding system read the motor commands instead of the moles, since there is a relation among moles, keys, motor commands, and letters? Maybe, but then how efficient is that? Why not just read the message of the moles directly from the moles themselves rather than routing it through the motor system?

Perhaps the power of predictive coding makes it worth the trouble. We route understanding through the motor system, one might argue, because *that* is how we can unleash a powerful computational mechanism. This argument loses force when we recognize that the motor system isn't the only source of predictive coding. Consider our GREEN EGGS AND HAM example. It is unlikely that our ability to predict AND HAM from GREEN EGGS stems from a motor familiarity associated with typing that sequence of letters. Rather it is a higher-level familiarity associated with understanding the name of a famous book.

Here's the question that worries me then; it should have a familiar ring. Why bother running the predictive process through the circuitous route of the motor system when we can use the mole reader itself to recognize the familiar pattern and generate an *understanding-based* forward mole whacking prediction that whacks moles before they appear (if the prediction is correct). It's the same computational mechanism as motor-generated predictive coding, with the same computational advantages, but now instead of detecting motor error, the system is detecting message error. If no moles pop it means the message reader already predicted the correct message. Done. If a mole pops, it means the prediction is wrong, the popped mole can then be read, and the prediction revised. It works just like predictive coding in the motor system but from a different predictive source. We might call it *ventral stream predictive coding,* to reflect the source and distinguish it from dorsal stream motor prediction.

Is there any evidence beyond the GREEN EGGS AND HAM example for nonmotor predictive coding? Indeed, there is a long history in psychology and neuroscience associated with the idea that perception is not simply a result of stimulation of the eyes, ears, and skin but an active *construction* of the brain. American psychologist William James distilled a "general law of perception" in his 1890 treatise *Principles of Psychology*:

> Whilst part of what we perceive comes through our senses from the object before us, another part (and it may be the larger part) always comes . . . out of our own head.[19]

The principle is illustrated in the figure here:

When you first see the image it typically doesn't look like much. But after a while you may notice the form of a dog—a Dalmatian in particular—right in the center of the image (if you can't find it, see page 254). The interesting thing is, once you see the Dalmatian, you can't "unsee" it. Your brain has developed an expectation, a (non-motor) prediction, for the object contained in the image and you can't help but see it. This effect is often referred to as an example of *top-down processing:* the use of higher-level information to inform lower-level sensory processing. The importance of top-down processing in perception is underlined by the neuroanatomy of sensory systems. Surprisingly, there are more top-down neural connections (from higher to lower levels) than bottom-up connections (from lower to higher levels), about an order of magnitude difference (10:1 ratio).[20]

Top-down processing may indeed be the larger part of perception, as James proposed.

Top-down processing of the sort illustrated by the Dalmatian image can be conceptualized as an example of ventral stream predictive coding. According to this view, a higher-level visual image, memory, or bit of contextual information (the mole reader) generates a prediction for lower-level sensory features in a stimulus (the pattern of the moles). The interaction of the predictive code with the sensory signals then either confirms the prediction (moles are whacked) or generates an error signal (moles pop), which would trigger higher levels to make the appropriate adjustments. In the Dalmatian example, viewing the image would trigger a visual memory of what's in the scene—that there is a dog, its size, location, and orientation—which would generate a prediction of where the contour lines, say, of the animal should appear. Some theorists have argued that indeed such top-down perceptual predictions are realized as suppression signals,[21] just like motor-generated predictions, although this hypothesis is still being assessed.

I'm suggesting that there are two sources of predictive coding, one from the dorsal stream motor system and the other from the ventral stream perceptual recognition system. It's a *dual-stream model of predictive coding*. Both streams use the same computational mechanism but for two different purposes, motor control and perceptual recognition. Because predictive coding was such an exciting discovery in the context of motor control, I think some theorists, myself included, rushed into the view that motor-generated predictions can enhance perception. I now believe that motor system and mirror neuron prediction operate squarely within the dorsal stream and play little role in perceptual recognition. But this is an empirical question. We'll have to wait and see what the data tell us.

———— • ————

MIRROR NEURONS are interesting creatures. On this there is agreement. The mirror neuron theory of action understanding is similarly interesting and, even if it is wrong, has served the field well by sparking a wealth of new findings, ideas, and counter ideas. On a personal level I can attest to how mirror neuron–inspired theories have pushed me to think harder about how language and other domains of cognition are implemented in the brain.

On a broader scale, mirror neurons have helped fuel a rethinking of how the mind works, how it is supported by that hunk of meat we call the brain, and how it evolved. When the monkey cells were discovered in the early 1990s the computational theory of mind was the dominant model. Mental life was viewed as a consequence of a set of modular, abstract, computational systems,[22] the antidote viewpoint to the behaviorism of Skinner, which dominated psychology during the first half of the twentieth century. Now the pendulum of opinion has swung away from that computational perspective toward a more environmentally grounded and embodied viewpoint. Mirror neurons played a major role in fostering that pendulum swing.

Although I don't believe that grounded/embodied theories succeeded in doing away with computation, I do believe that these pendulum swings are important for scientific progress. They push theorists to sharpen their arguments and experimentalists to design new studies to assess the validity of old and new claims. Pendulum swings and the theoretical push and pull that result create the environment for a kind of natural selection of scientific theories. Without environmental change, new species of ideas have no pressure to evolve and old species of ideas have no pressure to sharpen or fail.

I predict that mirror neurons will eventually be fully incorporated into a broad class of sensorimotor cells that participate in controlling action using a variety of sensory inputs and through a hierarchy of circuits. Motor cells that respond to actions in a nonmirror fashion (you-do-this, I-do-that)—which are about as prevalent as mirror neurons in monkey F5—will be as theoretically interesting in this respect as mirror neurons themselves. Action understanding will itself remain poorly understood, but its cognitive (computational) complexity will once again be appreciated; after all, given the number of moving parts involved, sorting out how action understanding works is darn close to solving the how-does-the-brain-work problem.

I also predict a gradual merging of the embodied cognition movement back into the folds of a more neurally "grounded" computational theory of mind. This will happen once we recognize that computation doesn't have to involve mathematical operators represented in the highest tiers of the mind but rather is achieved on scales great and small: by networks of neurons from microcircuits in the retina or spinal cord to massively interactive networks orchestrated by a neural CEO (Cerebral Executive Officer) in the prefrontal cortex. *Simulation* or *mirroring* will still, of course, be carried out by the brain, but we will in the future replace these terms with descriptions of the content of the computations for which they stand in.

The Parma group built a mirror neuron barn using only half of the lumber. They homed in on those cells that showed congruent execution-observation responses to the exclusion of cells that showed nonmirror (you-do-this, I-do-that) responses. They emphasized the centrality of the motor system in cognition to the exclusion of sensory systems, high-level multimodal systems, and the dual stream architecture of the brain. The result was a Hollywood set–style barn: it only looked good from one angle. Venture around back, or inside, and the fact that it is merely a façade becomes apparent. But parts of the façade, such as the action-related sensorimotor association features, are quite attractive and can be recycled in the barn renovation project currently underway in labs around the world. Placed in the context of a more balanced and complex structure, mirror neurons will no doubt have a role to play in our models of the neural basis of communication and cognition.

Appendix A

A Primer on Brain Organization

The brain can be divided into four main parts: *brain stem, diencephalon, cerebellum,* and *cerebrum* (see figure). The *brain stem* lies along the midline at the very base of the brain and houses areas important for controlling respiration, cardiovascular function, alertness, and other basic life-sustaining functions.

The *diencephalon* sits atop the brain stem in the center of the brain and contains two important structures, the *thalamus*, which acts like a switchboard connecting many different systems to one another, and the *hypothalamus*, which is important in controlling hormonal systems.

The *cerebellum* (literally "little brain" in Latin) is an important part of the motor system where it is involved in fine-tuning movements. Alcohol intoxication seems to impact cerebellar function significantly, so to get a sense of what the cerebellum does for motor control consider the effects of being drunk: staggering gait, slurred speech, clumsy movements. The cerebellum may also play a role in other functions such as language and attention. There are two cerebellar hemispheres.

The *cerebrum* is the large structure that sits on top of and surrounds all the other structures. It is the most visible part of the brain and the structure that is critical for sensory, motor, and other cognitive functions. The cerebrum does not work alone, however. It is densely connected to the other parts of the brain and interacts extensively

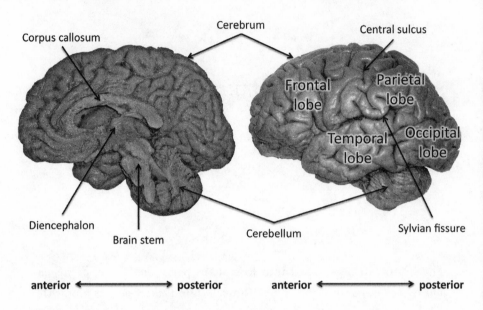

MEDIAL VIEW

Corpus callosum

Cerebrum

LATERAL VIEW

Central sulcus

Frontal lobe

Parietal lobe

Temporal lobe

Occipital lobe

Diencephalon

Brain stem

Cerebellum

Sylvian fissure

anterior ←——————→ posterior   anterior ←——————→ posterior

with them. There are two cerebral hemispheres, connected primarily by the corpus callosum. The surface of the cerebrum is composed of a three- to four-millimeter-thick sheet of neuron cell bodies, the cerebral cortex, or cortex for short. Virtually all of the discussion in this book concerns this part of the brain. The cortex is actually a layered structure. Most of the cortex contains six layers although evolutionarily older parts of the cortex in the limbic system (the portion of cortex involved in memory and emotion functions) contain three layers. Six-layered cortex is also referred to as *neocortex*. Beneath the cortex lies the *subcortex*. It contains a range of structures. Most of the volume of the subcortex comprises *white matter*, which are neuronal fiber tracts that connect neurons to one another. The *basal ganglia* circuit is the other major structure in the subcortex of the cerebrum. This system is also important for movement control, but rather than fine-tuning movements like the cerebellum, the basal ganglia are important for movement selection and initiation. This is the system that is affected in Parkinson's disease. The basal ganglia also play a role in nonmotor function.

We've known since the 1800s, primarily through neurological research on speech and language, that the cerebral cortex is not uniform in function. Different regions of the cortex do different things, but not in isolation. The cortex is kind of like the Internet: different nodes in the network do different things and store different kinds of information, but the system as a whole is widely interconnected.

With that caveat in mind we can outline some broad organizational features. The cortex can be roughly divided into three (highly interactive) parts: sensory systems, motor systems, and what I call *internal state* systems. The latter refers to memory and emotion-related networks, which rely on cortex mostly on the medial surface of the cerebral hemispheres. Sensory systems are found in the occipital, temporal, and parietal lobes, while motor systems are found in the frontal lobe. Again, this is a dramatic oversimplification but it does hold true for the *primary sensory and motor fields*, which brings us to the notion of cortical hierarchies.

The brain is an information processing device that operates on multiple levels of analysis, or degrees of granularity, which start at a low, fine-grained level and increase progressively as the information is processed in more depth. The visual system, for example, analyzes scenes by breaking them down into features and then putting the pieces back together. At the lowest level, the finest granularity, neurons in the retina code the presence or absence of spots of light or wavelength (color). Information from the retina is sent to the thalamus, where information from the two eyes is kept segregated, and then into the cerebral cortex, entering along a smallish swath of tissue on the medial surface of the occipital lobes called the *primary visual cortex*, or V1. This region codes visual information at a slightly higher level. Rather than spots of light, it codes for (analyzes) lines in particular orientations, edges, and movement direction; information from the two eyes is integrated. Beyond V1 are dozens of further visual areas covering much of the temporal and parietal lobes that integrate information across larger chunks of visual space and code ever more complex combinations of features. For instance, a well-known region several processing stages deep into the human cortical visual system

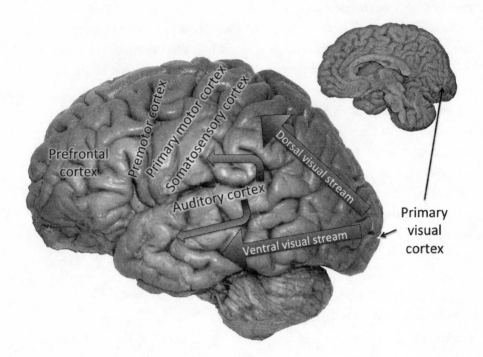

appears to code faces, the *fusiform face area*. The general picture of sensory processing, then, is almost assembly line–like: early (lower-level) stages of processing work on small pieces, while later stages build on the lower-level work until some unified whole emerges. Exactly what that unified whole looks like isn't entirely known and probably varies depending on what the brain is trying to do with the sensory information, as we'll see momentarily (and in various places in the present book).

Just like there is a primary visual cortex, other sensory system have their primary areas, or fields. In general, the portion of brain tissue where information from the sense organs first enters the cerebral cortex is called *primary visual, auditory,* or *somatosensory (body sense) cortex*, depending on which sensory modality one is talking about. The location of these primary cortical fields is shown in the figure above, except for primary auditory cortex, which is hidden in the depths of the Sylvian fissure.

The motor system is also organized hierarchically, but operates in reverse direction. *Primary motor cortex* (M1), located just in front of the central sulcus, is essentially the last stop out of the cerebral cortex when the brain executes a movement. Neurons in M1 code lower-level movement features like movement direction and send their signals down the spinal cord or out in the cranial nerves that link to muscles on the face or in the mouth. Motor processing stages prior to M1 occur in a region called *premotor cortex*, which is actually a collection of regions, but for our purposes we don't need to delve into the details. Premotor areas code movements at a higher level, such as sequences of movements or coordinated movements involving multiple joints or limbs. Mirror neurons were found initially in monkey premotor cortex. Human Broca's area, a prominent player in the present volume, is also probably a premotor-like area in the sense that it is involved in coding movements at a high level. However, the terminology can be confusing because Broca's area is not usually referred to as part of human premotor cortex. I simply refer to Broca's area as such.

In front of premotor cortex is *prefrontal cortex*, the region of the brain that supports the highest-level cognitive abilities: reasoning, planning, decision making, and other so-called "executive" functions. We do not venture too deeply into the mysterious prefrontal cortex in this book.

These organizational generalizations should be viewed as anchor points for understanding cerebral cortex function, not as immutable facts. It's worth reemphasizing that these systems interact extensively, often to the point where it is meaningless to draw any boundaries at all between systems. For example, a common view is that when we see an object, the visual cortex processes the visual information, recognizes the object, and then hands off its handiwork to other systems to remember the object, or reach for the object, or talk about the object. But this isn't how sensory processing works. In fact there isn't just a single visual system, there are (at least) two, one for interacting with the motor system, the *dorsal stream*, and one for interacting with the conceptual memory system, the *ventral stream*. The names of the streams derive from their physical locations in the brain; one is more dorsally located (toward the top of the brain) and the other more ventral (toward the bottom). The dorsal stream plays an important role in sensorimotor functions (using sensory information to guide action) whereas the ventral stream is responsible for recognizing the content of visual information (linking it to previous experiences with similar objects). There is strong evidence for dual-stream architectures in both vision and hearing (indicated by the arrows in the figure), and some evidence for the same in the somatosensory system. The dual-stream architecture is discussed in various places in this book.

Appendix B

Cognitive Neuroscience Toolbox

Resolution issues Every brain research method has its strengths and weaknesses. A major consideration in this respect is *resolution*, which can vary in the spatial and temporal dimensions. Spatial resolution refers to the *where* question: how precisely we determine where the signal is coming from. Temporal resolution refers to the *when* question: how precisely we can determine when the signal was generated. To illustrate, imagine two different smartphone tracking apps that you are considering for installation on your child's phone. App 1 generates a location signal every second, but the location is accurate to within a kilometer only. App 2 generates a signal every hour but is accurate to within a meter. If you choose App 1 you will know only roughly where the phone is (poor spatial resolution) but if there is a (big) change in location, you will know it right away (good temporal resolution). If you choose App 2 you will know precisely where the phone is when the signal was generated (good spatial resolution), but if the phone moves, you won't know for a long time and you'll never see movement that occurs between signal broadcasts (poor temporal resolution). Of course you would prefer App 3, which generates signals every second with meter precision. Neuroscientists are in the same position. We want a measurement tool that has both high spatial and temporal resolution, but unfortunately we often have to choose one or the other.

Lesion method   The oldest method but still highly valuable for analyzing the brain regions involved in cognition. The approach correlates damage to various brain regions with specific behavioral deficits. The benefit of the lesion method is that it is instructive in assessing causal relations between brain and behavior. If the ability to speak is severely disrupted following brain damage to area x, then we can infer that area x is critically involved in *some aspect* of the process of speaking.

EEG   Electroencephalography or EEG measures the electrical activity of large populations of neurons using surface electrodes attached to the scalp. It has good temporal resolution (samples are taken every millisecond) but relatively poor spatial resolution. The reason for the poor spatial resolution is twofold. One is that the EEG signal is generated by large populations of cells that can be distributed over moderately widespread brain areas. The other is that electric signals coming from different populations of neurons intermix and get smeared as the current travels to the scalp, finding the electrical path of least resistance, which makes it difficult to identify the source of the signal(s). In patients undergoing brain surgery, EEG recordings can be made directly on the surface of the brain, a technique called ECoG for electrocorticography, which enhances source localization and therefore provides both high temporal and spatial resolution.

MEG   Magnetoencephalography or MEG measures the magnetic fields associated with the electrical activity of large populations of neurons using sensor arrays situated around the head (much like an old-fashioned salon hairdryer). MEG, like EEG, has millisecond temporal resolution, and because magnetic fields are not affected by the conductance properties of tissue that they pass through it avoids one of the complications of localization inherent in EEG. However, signal intermixing is still a significant problem in MEG.

PET Positron emission tomography or PET is a *hemodynamic* imaging method. It measures brain activity indirectly via changes in blood flow. When neurons are active, blood flow to the active brain region increases locally to satisfy the increased metabolic demand from the neurons. PET measures these changes in blood flow in a straightforward manner. In a typical measurement, water is labeled with a radioactive tracer, oxygen-15, and introduced into the blood stream; metabolically active regions in the brain receive a greater concentration of the radioactive tracer. Regional concentrations of the tracer can then be measured in the brain by using a PET scanner, which detects the decay of the tracer. PET has decent spatial resolution (around one centimeter) but very poor temporal resolution, taking measurements on the scale of minutes. PET has largely been replaced by fMRI but is useful for researchers to label different compounds, such as glucose or certain neurotransmitters, which allow for different measurements.

fMRI Functional magnetic resonance imaging, or fMRI, measures roughly the same blood flow changes as PET, but it does so in a different way. Changes in blood flow result in changes in the oxygen concentration of the blood, which in turn changes the signal intensity on a normal MRI image. By taking lots of MR images rapidly in time, rather than just one to get a single pretty image, fMRI can pick up *changes* in the image over time. These changes correlate with changes in the oxygen concentration of the blood that correlate with changes in blood flow that correlate with neural activity. While it is not the most direct measure, fMRI has served the field well by providing very high spatial resolution (down to a few millimeters) and so-so temporal resolution (on the order of seconds).

Functional brain imaging issues Because fMRI and PET are effectively imaging neural activity, or function, rather than brain structure, they are referred to as "functional brain imaging" methods. Despite the clean, pretty pictures that we see in journal articles or in the press, it is important to recognize that the data behind those pictures are rather noisy and can be imprecise. One major complication comes from the fact that brains vary from person to person both anatomically and functionally. It is helpful to think of brains as being similar to faces: they all have the same basic structure at a macro level but vary in the details. This is not a problem if the goal of a study is to image the anatomy and map the function of an individual. However, if we want to make inferences about how "the brain" works in general, we need to sample a group of subjects and then somehow average their anatomy and functional activation patterns. The typical approach in functional imaging is to normalize (essentially warp) the digital brain images of individuals into a common anatomical space and then average the functional activity across subjects to identify which patterns are consistently observed in the group. There are two sources of error associated with this process. One is that the anatomy of the individuals won't line up perfectly and the other is that the functional boundaries within the anatomical structures won't line up perfectly. The effect of this error is that the spatial resolution of group-averaged brain imaging data is worse than the spatial resolution that a method is capable of in individual subjects. Sometimes this can lead to mislocalization. Imagine averaging (morphing) two faces, one with wide-set eyes and the other with narrow-set eyes. The average of the two would place the eyes between the extremes, a location that does not represent the true location in either individual. Averaging can also decrease sensitivity. Many people have blemishes on their faces, but if you average a group of faces, the blemishes tend to vanish because they are not in the same place on different faces. The same is true in brain activation patterns. If a functional area is relatively small (say less than a square centimeter) and its location varies by more than a centimeter from person to person, it tends to vanish in the average. For this reason, some neuroimaging researchers avoid group averaging in favor of individual subject analyses. This is common in mapping visual cortex, for example, where the functional areas are small and variable. While extremely powerful, functional imaging is not a magic bullet method for revealing the truth about the organization of the brain.

TMS Transcranial magnetic stimulation or TMS doesn't measure anything. Instead it deposits energy via a strong magnetic pulse that in turn generates electrical current in the brain to activate the tissue it stimulates or to interrupt function temporarily. The spatial resolution of TMS is on the order of a centimeter.

ASIDE FROM the lesion method, which typically involves stroke patients, and ECoG, which involves patients undergoing brain surgery for clinical reasons (usually epilepsy or tumors), all of these methods are considered noninvasive and are used in healthy populations, which allows us to study the brain in its normal state. Because of the radioactive material used in taking PET measurements, there are strict guidelines on how many studies a participant can be involved in (much like an x-ray).

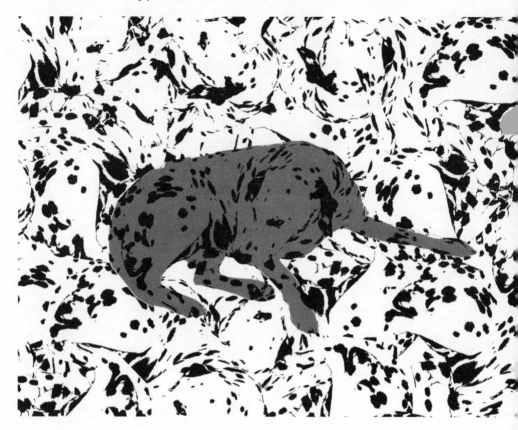

The Dalmatian revealed, from page 238

Acknowledgments

TO BE perfectly honest and to skirt the cliché, I probably *could* have finished this book without anyone's help. However, it would never have been published and probably wouldn't be worth reading anyway. I am, therefore, deeply thankful and indebted to the many people who provided advice, encouragement, commentary, criticism, artistic and editorial assistance.

A number of friends and colleagues provided critical feedback on the content of various chapters, including Michael Arbib, Patricia Churchland, Steve Cramer, Morton Ann Gernsbacher, Melvyn Goodale, Cecelia Heyes, Norbert Hornstein, and Kourosh Saberi. My longtime friend and collaborator David Poeppel read and commented on large sections of the book and provided the occasional, much needed general encouragement. I am especially indebted to Steve Pinker who provided practical advice at several stages of the project, including a gripping tutorial on the use of the comma (I still don't understand them). He also fielded random questions about *how the mind works* (he wrote the book after all: S. Pinker [1997], Norton), read and commented on portions of the manuscript, and helped with the title.

My daughters, Taylor and Ally, provided invaluable help on the manuscript. Thank you, Taylor, for your editorial feedback and corrections. Thank you, Ally, for lending me your artistic and computer graphic talents; the figure credits are all yours.

I'd also like to thank Sandy Blakeslee for providing early encouragement on the book idea and connecting me with her agent, Jim Levine, who is now also my agent. Jim and his team provided invaluable advice and commentary on the development of the book, for which I am deeply grateful. My editor, Brendan Curry, deserves many thanks not only from me, but from you, the reader. His editing dramatically improved the readability and pushed me to develop some of my ideas more fully. And Carol Rose did a fabulous job at the copyedit stage.

Going back a bit further, it is important to acknowledge the many students, colleagues, reviewers, journal editors, audience members, and bloggers who played a role in sharpening my ideas on mirror neurons and the neural basis of communication and cognition. There are too many to mention but I will single out a few. Mark D'Esposito and Rich Ivry were instrumental in getting my 2009 mirror neuron critique published in the *Journal of Cognitive Neuroscience*, which laid the groundwork for this book; I appreciate their editorial savvy. E-mail and face-to-face discussions with Michael Arbib, Vittorio Gallese, Art Glenberg, Luciano Fadiga, and Corrado Sinigaglia were key in helping clarify various perspectives on mirror neuron function. Michael Arbib in particular was quite helpful in pointing out many papers and viewpoints that I had overlooked. Finally, although I don't know all of their names, I am grateful to the many individuals who read and sometimes anonymously commented on my mirror neuron–related posts on the *Talking Brains* blog. They were fun and instructive exchanges.

Going all the way back, I would like to thank my parents for their endless support. Unlike all the others mentioned above, I truly couldn't have done it without you.

Most of all, I want to thank my wife, Krissy. For someone who claims not to work in my field, you provided an amazing amount of substantive feedback as we walked through my ideas and arguments. More than that, though, I cherish your constant encouragement and support. You are my spark.

Notes

PREFACE: *A Neural Blueprint for Human Behavior?*

1   Watson, J. D., & Crick, F. H. (1953). Molecular structure of nucleic acids; a structure for deoxyribose nucleic acid. *Nature, 171,* 737–738.

2   Ramachandran, V. S. (2000). MIRROR NEURONS and imitation learning as the driving force behind "the great leap forward" in human evolution. *Edge.org.* http://www.edge.org/3rd_culture/ramachandran/ramachandran_index.html%3E.

3   Iacoboni, M. (2008). *Mirroring people: The new science of how we connect with others.* New York: Farrar, Straus and Giroux.

4   Mouras, H., et al. (2008). Activation of mirror-neuron system by erotic video clips predicts degree of induced erection: An fMRI study. *Neuroimage, 42,* 1142–1150, doi:10.1016/j.neuroimage.2008.05.051.

5   Rizzolatti, G., & Craighero, L. (2004). The mirror-neuron system. *Annual Review of Neuroscience, 27,* 169–192.

CHAPTER 1: *Serendipity in Parma*

1   John Eccles in 1963 for his work on motor neurons and David Hubel and Torsten Weisel in 1981 for their work on the visual system.

2   Rizzolatti, G., et al. (1988). Functional organization of inferior area 6 in the macaque monkey. II. Area F5 and the control of distal movements. *Experimental Brain Research. Experimentelle Hirnforschung. Experimentation Cerebrale, 71,* 491–507.

3   di Pellegrino, G., Fadiga, L., Fogassi, L., Gallese, V., & Rizzolatti, G. (1992). Understanding motor events: A neurophysiological study. *Experimental Brain Research, 91,* 176–180.

4   Perrett, D. I., et al. (1985). Visual analysis of body movements by neurones in the temporal cortex of the macaque monkey: A preliminary report. *Behavioral Brain Research, 16,* 153–170.

5   di Pellegrino, Fadiga, Fogassi, Gallese, & Rizzolatti. (1992). Understanding motor events.

6   Heilman, K. M., Rothi, L. J., & Valenstein, E. (1982).Two forms of ideomotor apraxia. *Neurology, 32,* 342–346.

7   Broca, P. (1861). Remarques sur le siège de la faculté du langage articulé; suivies d'une observation d'aphémie (perte de la parole). *Bulletins de la Société Anatomique (Paris), 6,* 330–357, 398–407; and Broca, P. (1865). Sur le siège de la faculté du langage articulé. *Bulletins de la Société d'Anthropologie, 6,* 337–393.

8   Carey, D. P. (1996). "Monkey see, monkey do" cells. Neurophysiology. *Current Biology, 6,* 1087–1088.

9   Gallese, V., Fadiga, L., Fogassi, L., & Rizzolatti, G. (1996). Action recognition in the premotor cortex. *Brain, 119 (Pt 2),* 593–609.

10   Fadiga, L., Fogassi, L., Pavesi, G., & Rizzolatti, G. (1995). Motor facilitation during action observation: A magnetic stimulation study. *Journal of Neurophysiology, 73,* 2608–2611.

11   Rizzolatti, G., et al. (1996). Localization of grasp representations in humans by PET: 1. Observation versus execution. *Experimental Brain Research, 111,* 246–252.

CHAPTER 2: *Like What DNA Did for Biology*

1   Rizzolatti & Craighero (2004). The mirror-neuron system.

2   Rizzolatti, G., & Arbib, M. (1998). Language within our grasp. *Trends in Neurosciences, 21,* 188–194.

3   Gallese, V., & Goldman, A. (1998). Mirror neurons and the simulation theory of mind-reading. *Trends in Cognitive Sciences, 2,* 493–501.

4   Gallese, V. (2001). The "shared manifold" hypothesis: From mirror neurons to empathy. *Journal of Consciousness Studies, 8,* 33–50; quotation appears on p. 44.

5   Williams, J. H., Whiten, A., Suddendorf, T., & Perrett, D. I. (2001). Imitation, mirror neurons and autism. *Neuroscience and Biobehavioral Reviews, 25,* 287–295.

6   Nishitani, N., & Hari, R. (2002). Viewing lip forms: Cortical dynamics. *Neuron, 36,* 1211–1220.

7   Kalinowski, J. & Saltuklaroglu, T. (2003). Speaking with a mirror: Engagement of mirror neurons via choral speech and its derivatives induces stuttering inhibition. *Medical Hypotheses, 60,* 538–543.

8   Fahim, C., et al. (2004). Negative socio-emotional resonance in schizophrenia: A functional magnetic resonance imaging hypothesis. *Medical Hypotheses, 63,* 467–475. doi:10.1016/j.mehy.2004.01.035.

9   Hauk, O., Johnsrude, I., & Pulvermuller, F. (2004). Somatotopic representation of action words in human motor and premotor cortex. *Neuron, 41,* 301–307.

10   Iacoboni, M. (2005). Neural mechanisms of imitation. *Current Opinion in Neurobiology, 15,* 632–637. doi:10.1016/j.conb.2005.10.010.

11   Price, E. H. (2006). A critical review of congenital phantom limb cases and a developmental theory for the basis of body image. *Consciousness and Cognition, 15,* 310–322. doi:10.1016/j.concog.2005.07.003.

12 Buccino, G., Solodkin, A., & Small, S. L. (2006). Functions of the mirror neuron system: Implications for neurorehabilitation. *Cognitive and Behavioral Neurology: Official Journal of the Society for Behavioral and Cognitive Neurology, 19*, 55–63.

13 Rossi, E. L., & Rossi, K. L. (2006). The neuroscience of observing consciousness & mirror neurons in therapeutic hypnosis. *The American Journal of Clinical Hypnosis, 48*, 263–278.

14 Gridley, M. C., & Hoff, R. (2006). Do mirror neurons explain misattribution of emotions in music? *Perceptual and Motor Skills, 102*, 600–602.

15 Ponseti, J., et al. (2006). A functional endophenotype for sexual orientation in humans. *Neuroimage, 33*, 825–833. doi:10.1016/j.neuroimage.2006.08.002.

16 Pineda, J. O., & Oberman, L. M. (2006). What goads cigarette smokers to smoke? Neural adaptation and the mirror neuron system. *Brain Research, 1121*, 128–135. doi:10.1016/j.brainres.2006.08.128.

17 Molnar-Szakacs, I., & Overy, K. (2006). Music and mirror neurons: From motion to "e"motion. *Social Cognitive and Affective Neuroscience, 1*, 235–241. doi:10.1093/scan/nsl029.

18 Kaplan, J. T., Freedman, J., & Iacoboni, M. (2007). Us versus them: Political attitudes and party affiliation influence neural response to faces of presidential candidates. *Neuropsychologia, 45*, 55–64. doi:10.1016/j.neuropsychologia.2006.04.024.

19 Nielsen, T. (2007). Felt presence: paranoid delusion or hallucinatory social imagery? *Consciousness and Cognition, 16*, 975–983; discussion 984–991. doi:10.1016/j.concog.2007.02.002.

20 Enticott, P. G., Johnston, P. J., Herring, S. E., Hoy, K. E., & Fitzgerald, P. B. (2008). Mirror neuron activation is associated with facial emotion processing. *Neuropsychologia, 46*, 2851–2854. doi:10.1016/j.neuropsychologia.2008.04.022.

21 Cohen, D. A. (2008). Neurophysiological pathways to obesity: Below awareness and beyond individual control. *Diabetes, 57*, 1768–1773. doi:10.2337/db08-0163.

22 Mouras et al. (2008). Activation of the mirror-neuron system by erotic video clips.

23 Fecteau, S., Pascual-Leone, A., & Theoret, H. (2008). Psychopathy and the mirror neuron system: Preliminary findings from a non-psychiatric sample. *Psychiatry Research, 160*, 137–144. doi:10.1016/j.psychres.2007.08.022.

24 Ortigue, S., & Bianchi-Demicheli, F. (2008). Why is your spouse so predictable? Connecting mirror neuron system and self-expansion model of love. *Medical Hypotheses, 71*, 941–944. doi:10.1016/j.mehy.2008.07.016.

25 Cooper, N. R., Puzzo, I., & Pawley, A. D. Contagious yawning: The mirror neuron system may be a candidate physiological mechanism. *Medical Hypotheses, 71*, 975–976. doi:10.1016/j.mehy.2008.07.023.

26 Goleman, D., & Boyatzis, R. (2008). Social intelligence and the biology of leadership. *Harvard Business Review, 86*, 74–81, 136.

27 Moriguchi, Y., et al. (2009). The human mirror neuron system in a population with deficient self-awareness: An fMRI study in alexithymia. *Human Brain Mapping, 30*, 2063–2076. doi:10.1002/hbm.20653.

28 Diamond, D. (2008). Empathy and identification in Von Donnersmarck's "The Lives of Others." *Journal of the American Psychoanalytic Association, 56,* 811–832. doi:10.1177/0003065108323590.

29 Lenzi, D., et al. (2009). Neural basis of maternal communication and emotional expression processing during infant preverbal stage. *Cerebral Cortex, 19,* 1124–1133. doi:10.1093/cercor/bhn153.

30 Ramachandra, V., Depalma, N., & Lisiewski, S. (2009). The role of mirror neurons in processing vocal emotions: Evidence from psychophysiological data. *The International Journal of Neuroscience, 119,* 681–690. doi:10.1080/00207450802572188.

31 Shimada, S., & Abe, R. (2009). Modulation of the motor area activity during observation of a competitive game. *Neuroreport, 20,* 979–983. doi:10.1097/WNR.0b013e32832d2d36.

32 Knox, J. (2009). Mirror neurons and embodied simulation in the development of archetypes and self-agency. *The Journal of Analytical Psychology, 54,* 307–323. doi:10.1111/j.1468-5922.2009.01782.x; and Hogenson, G. B. (2009). Archetypes as action patterns. *The Journal of Analytical Psychology, 54,* 325–337. doi:10.1111/j.1468-5922.2009.01783.x.

33 Lee, Y. T., & Tsai, S. J. (2010). The mirror neuron system may play a role in the pathogenesis of mass hysteria. *Medical Hypotheses, 74,* 244-245, doi:10.1016/j.mehy.2009.09.031.

34 Newlin, D. B., & Renton, R. M. (2010). A self in the mirror: mirror neurons, self-referential processing, and substance use disorders. *Substance Use & Misuse, 45,* 1697–1726. doi:10.3109/10826084.2010.482421.

35 Kourtis, D., Sebanz, N., & Knoblich, G. (2010). Favouritism in the motor system: Social interaction modulates action simulation. *Biology Letters, 6,* 758–761. doi:10.1098/rsbl.2010.0478.

36 Botbol, M. (2010). Towards an integrative neuroscientific and psychodynamic approach to the transmission of attachment. *Journal of Physiology, Paris, 104,* 263–271. doi:10.1016/j.jphysparis.2010.08.005.

37 Schermer, V. L. (2010). Mirror neurons: Their implications for group psychotherapy. *International Journal of Group Psychotherapy, 60,* 486–513. doi:10.1521/ijgp.2010.60.4.486.

38 Blanchard, D. C., Griebel, G., Pobbe, R., & Blanchard, R. J. (2011). Risk assessment as an evolved threat detection and analysis process. *Neuroscience and Biobehavioral Reviews, 35,* 991–998. doi:10.1016/j.neubiorev.2010.10.016.

39 Saurat, M. T., Agbakou, M., Attigui, P., Golmard, J. L., & Arnulf, I. (2011). Walking dreams in congenital and acquired paraplegia. *Consciousness and Cognition, 20,* 1425–1432. doi:10.1016/j.concog.2011.05.015.

40 Fitzgibbon, B. M., et al. (2011). Enhanced corticospinal response to observed pain in pain synesthetes. *Cognitive, Affective & Behavioral Neuroscience.* doi:10.3758/s13415-011-0080-8.

41 Herman, L. M. (2012). Body and self in dolphins. *Consciousness and Cognition, 21,* 526–545. doi:10.1016/j.concog.2011.10.005.

CHAPTER 3: *Human See, Human Do?*

1 Fadiga, Fogassi, Pavesi, & Rizzolatti (1995), Motor facilitation during action observation.

2 Grezes, J., Tucker, M., Armony, J., Ellis, R., & Passingham, R. E. (2003). Objects automatically potentiate action: An fMRI study of implicit processing. *European Journal of Neuroscience, 17,* 2735–2740.

3 Rizzolatti et al. (1996), Localization of grasp representations in humans by PET: 1.

4 Rizzolatti et al. (1996), Localization of grasp representations in humans by PET: 1.

5 Perrett et al. (1985), Visual analysis of body movements by neurones; and Perrett, D. I., Mistlin, A. J., Harries, M. H., & Chitty, A. J. (1990). Understanding the visual appearance of hand actions. In M. A. Goodale (Ed.), *Vision and action: The control of grasping* (pp. 163–180). New York: Ablex.

6 Grafton, S. T., Arbib, M. A., Fadiga, L., & Rizzolatti, G. (1996). Localization of grasp representations in humans by positron emission tomography. 2. Observation compared with imagination. *Experimental Brain Research. Experimentelle Hirnforschung. Experimentation Cerebrale, 112,* 103–111.

7 Rizzolatti et al. (1996), Localization of grasp representations in humans by PET: 1, 251.

8 Binkofski, F., et al. (1999). A parieto-premotor network for object manipulation: Evidence from neuroimaging. *Experimental Brain Research. Experimentelle Hirnforschung. Experimentation Cerebrale, 128,* 210–213.

9 Grezes, J., & Decety, J. (2001). Functional anatomy of execution, mental simulation, observation, and verb generation of actions: A meta-analysis. *Human Brain Mapping, 12,* 1–19 .

10 Gallese, V., Fogassi, L., Fadiga, L., & Rizzolatti, G. (2002). Action representation and the inferior parietal lobule. In W. Prinz & B. Hommel (Eds.), *Common mechanisms in perception and action* (pp. 247–266). Oxford: Oxford University Press.

11 Iacoboni, M., et al. (1999). Cortical mechanisms of human imitation. *Science, 286,* 2526–2528.

12 Dinstein, I., Thomas, C., Behrmann, M., & Heeger, D. J. (2008). A mirror up to nature. *Current Biology, 18,* R13–18.

13 Dinstein, I., Hasson, U., Rubin, N., & Heeger, D. J. (2007). Brain areas selective for both observed and executed movements. *Journal of Neurophysiology, 98,* 1415–1427; Lingnau, A., Gesierich, B., & Caramazza, A. (2009). Asymmetric fMRI adaptation reveals no evidence for mirror neurons in humans. *Proceedings of the National Academy of Sciences of the United States of America, 106,* 9925–9930, doi:0902262106 [pii] 10.1073/pnas.0902262106; and Dinstein, I., Gardner, J. L., Jazayeri, M., & Heeger, D. J. (2008). Executed and observed movements have different distributed representations in human aIPS. *Journal of Neuroscience, 28,* 11231–11239, doi:10.1523/JNEUROSCI.3585-08.2008.

14   Kilner, J. M., Neal, A., Weiskopf, N., Friston, K. J., & Frith, C. D. (2009). Evidence of mirror neurons in human inferior frontal gyrus. *Journal of Neuroscience, 29*, 10153–10159. doi:10.1523/JNEUROSCI.2668-09.2009.

CHAPTER 4: *Anomalies*

1   Rizzolatti & Craighero (2004), *The mirror-neuron system*, 169–192.

2   Rizzolatti, G., & Sinigaglia, C. (2010). The functional role of the parieto-frontal mirror circuit: Interpretations and misinterpretations. *Nature Reviews Neuroscience, 11*, 264–274. doi:10.1038/nrn2805.

3   Buccino, G., et al. (2004). Neural circuits involved in the recognition of actions performed by nonconspecifics: An fMRI study. *Journal of Cognitive Neuroscience, 16*, 114–126.

4   Molnar, C., Pongracz, P., & Miklosi, A. (2010). Seeing with ears: Sightless humans' perception of dog bark provides a test for structural rules in vocal communication. *Quarterly Journal of Experimental Psychology (Hove), 63*, 1004–1013. doi:10.1080/17470210903168243.

5   Pazzaglia, M., Smania, N., Corato, E., & Aglioti, S. M. (2008). Neural underpinnings of gesture discrimination in patients with limb apraxia. *Journal of Neuroscience, 28*, 3030–3041.

6   Ekman, P., Sorenson, E. R., & Friesen, W. V. (1969). Pan-cultural elements in facial displays of emotion. *Science, 164*, 86–88; and Ekman, P., & Friesen, W. V. (1986). A new pan-cultural facila express of emotion. *Motivation and Emotion, 10*, 159–168.

7   Ekman, P., Levenson, R. W., & Friesen, W. V. (1983). Autonomic nervous system activity distinguishes among emotions. *Science, 221*, 1208–1210.

8   Bogart, K. R., & Matsumoto, D. (2010). Living with moebius syndrome: Adjustment, social competence, and satisfaction with life. *Cleft Palate-Craniofacial Journal, 47*, 134–142, doi:10.1597/08-257.1.

9   Bogart, K. R., & Matsumoto, D. (2010). Facial mimicry is not necessary to recognize emotion: Facial expression recognition by people with Moebius syndrome. *Social Neuroscience, 5*, 241–251.

10   Buonomano, D. V., & Merzenich, M. M. (1998). Cortical plasticity: From synapses to maps. *Annual Review of Neuroscience, 21*, 149–186. doi:10.1146/annurev.neuro.21.1.149.

11   Sharma, A., Dorman, M. F., & Spahr, A. J. (2002). A sensitive period for the development of the central auditory system in children with cochlear implants: Implications for age of implantation. *Ear & Hearing, 23*, 532–539, doi:10.1097/01.AUD.0000042223.62381.01.

12   Osberger, M. J., Zimmerman-Phillips, S., & Koch, D. B. (2002). Cochlear implant candidacy and performance trends in children. *Annals of Otology, Rhinology, and Laryngology, 189*, 62–65.

13   Zwolan, T. A., Kileny, P. R., & Telian, S. A. (1996). Self-report of cochlear implant use and satisfaction by prelingually deafened adults. *Ear and Hearing, 17*, 198–210.

14   Catmur, C., Walsh, V., & Heyes, C. (2007). Sensorimotor learning configures the human mirror system. *Current Biology, 17,* 1527–1531.

15   Ferrari, P. F., Rozzi, S., & Fogassi, L. (2005). Mirror neurons responding to observation of actions made with tools in monkey ventral premotor cortex. *Journal of Cognitive Neuroscience, 17,* 212–226; quotation from p. 213.

16   Milner, A. D., & Goodale, M. A. (1995). *The visual brain in action.* Oxford: Oxford University Press.

17   Milner & Goodale (1995), *The visual brain in action*; Hickok, G., & Poeppel, D. (2007). The cortical organization of speech processing. *Nature Reviews Neuroscience, 8,* 393–402; and Rauschecker, J. P. (1998). Cortical processing of complex sounds. *Current Opinion in Neurobiology, 8,* 516–521.

18   Farah, M. J. (2004). *Visual agnosia* (2nd ed). Cambridge, MA: MIT Press.

19   Milner & Goodale (1995), *The visual brain in action.*

20   Rossetti, Y., Pisella, L., & Vighetto, A. (2003). Optic ataxia revisited: Visually guided action versus immediate visuomotor control. *Experimental Brain Research. Experimentelle Hirnforschung. Experimentation Cerebrale, 153,* 171–179. doi:10.1007/s00221-003-1590-6.

21   Milner & Goodale, *The visual brain in action*; and Goodale, M. A., & Milner, A. D. (2004). *Sight unseen: An exploration of conscious and unconscious vision.* Oxford: Oxford University Press.

22   Buchsbaum, B. R., et al. (2011). Conduction aphasia, sensory-motor integration, and phonological short-term memory—an aggregate analysis of lesion and fMRI data. *Brain and Language, 119,* 119–128, doi:10.1016/j.bandl.2010.12.001; and Hickok, G. (2000). Speech perception, conduction aphasia, and the functional neuroanatomy of language. In Y. Grodzinsky, L. Shapiro, & D. Swinney (Eds.), *Language and the brain* (pp. 87–104). Waltham, MA: Academic Press.

23   Rizzolatti et al. (1988), Functional organization of inferior area 6 . . . : II.

24   Jeannerod, M., Arbib, M. A., Rizzolatti, G., & Sakata, H. (1995). Grasping objects: The cortical mechanisms of visuomotor transformation. *Trends in Neurosciences, 18,* 314–320. doi:016622369593921J [pii].

25   Jeannerod, Arbib, Rizzolatti, & Sakata (1995), Grasping objects.

26   Ferrari, Rozzi, & Fogassi (2005), Mirror neurons responding to observation, 212.

27   Gallese, Fadiga, Fogassi, & Rizzolatti (1996), Action recognition in the premotor cortex, 606.

28   Rizzolatti & Craighero (2004), The mirror-neuron system, 172.

29   Csibra, G. (2007). Action mirroring and action understanding: An alternative Account. In P. Haggard, Y. Rosetti, & M. Kawato (Eds.), *Sensorimotor foundations of higher cognition. Attention and performance XXII* (pp. 453–459). Oxford: Oxford University Press.

30   Fogassi, L., et al. (2005). Parietal lobe: From action organization to intention understanding. *Science, 308,* 662–667.

31   Umilta, M. A., et al. (2008). When pliers become fingers in the monkey motor system. *Proceedings of the National Academy of Sciences of the United States of America, 105,* 2209–2213. doi:10.1073/pnas.0705985105.

32 Iacoboni, M., et al. (2005). Grasping the intentions of others with one's own mirror neuron system. *PLoS Biology, 3,* e79.

33 Churchland, P. S. 2011. *Braintrust: What neuroscience tells us about morality.* Princeton University Press, Princeton, N.J.

34 Csibra (2007), in *Sensorimotor foundations of higher cognition,* 446–447.

35 Umiltà, M., et al. (2001). I know what you are doing. A neurophysiological study. *Neuron, 31,* 155–165.

36 This wasn't true for all mirror neurons. In fact, a little over half of them (18 of 35; I'm excluding data from the "placing" condition in which the object was visible during the first half of the action) did not respond in the hidden condition. The authors, of course, emphasized the positive results of the other 17 neurons that did respond and argued that mirror neurons "know what you are doing." But we could just as easily emphasize the reverse result and conclude that half of all mirror neurons are not coding the goal or meaning of the action. And if half the population of mirror neurons are not coding the goal of an action, then what are they coding when they fire? Then we should ask, does the answer to this question change our interpretation of mirror neurons generally?

37 Rizzolatti & Sinigaglia (2010), The functional role of the parieto-frontal mirror circuit.

38 Nelissen, K., Luppino, G., Vanduffel, W., Rizzolatti, G., & Orban, G. A. (2005). Observing others: Multiple action representation in the frontal lobe. *Science, 310,* 332–336.

39 Nelissen, Luppino, Vanduffel, Rizzolatti, & Orban (2005), Observing others.

40 Gallese, Fadiga, Fogassi, & Rizzolatti (1996), Action recognition in the premotor cortex.

CHAPTER 5: *Talking Brains*

1 Liberman, A. M. (1957). Some results of research on speech perception. *Journal of the Acoustical Society of America, 29,* 117–123; quotation from p. 117.

2 Harris, C. M. (1953). A study of the building blocks in speech. *Journal of the Acoustical Society of America, 25,* 962–969.

3 Liberman, A. M., Cooper, F. S., Shankweiler, D. P., & Studdert-Kennedy, M. (1967). Perception of the speech code. *Psychological Review, 74,* 431–461; quotation from p. 436.

4 Liberman, A. M., Harris, K. S., Hoffman, H. S., & Griffith, B. C. (1957). The discrimination of speech sounds within and across phoneme boundaries. *Journal of Experimental Psychology, 54,* 358–368.

5 Liberman, A. M., & Mattingly, I. G. (1989). A specialization for speech perception. *Science, 243,* 489–494; and Rand, T. C. (1974). Dichotic release from masking for speech. *Journal of the Acoustical Society of America, 55,* 678–680.

6 McGurk, H., & MacDonald, J. (1976). Hearing lips and seeing voices. *Nature, 264,* 746–748.

7 Eimas, P. D., Siqueland, E. R., Jusczyk, P., & Vigorito, J. (1971). Speech per-

ception in infants. *Science, 171,* 303–306; and Kuhl, P. K., & Miller, J. D. (1975). Speech perception by the chinchilla: Voiced-voiceless distinction in alveolar plosive consonants. *Science, 190,* 69–72.

8 Etcoff, N. L., & Magee, J. J. (1992). Categorical perception of facial expressions. *Cognition, 44,* 227–240; Nelson, D. A., & Marler, P. (1989). Categorical perception of a natural stimulus continuum: Birdsong. *Science, 244,* 976–978; Roberson, D., & Davidoff, J. (2000). The categorical perception of colors and facial expressions: The effect of verbal interference. *Memory & Cognition, 28,* 977–986; Wyttenbach, R. A., May, M. L., & Hoy, R. R. (1996). Categorical perception of sound frequency by crickets. *Science, 273,* 1542–1544; Locke, S., & Kellar, L. (1973). Categorical perception in a non-linguistic mode. *Cortex, 9,* 355–369; and Fowler, C. A., & Rosenblum, L. D. (1990). Duplex perception: A comparison of monosyllables and slamming doors. *Journal of Experimental Psychology. Human Perception and Performance, 16,* 742–754.

9 Massaro, D. W., & Cohen, M. M. (1983). Categorical or continuous speech perception: A new test. *Speech Communication, 2,* 15–35.

10 Massaro, D. W., & Stork, D. G. (1998). Speech recognition and sensory integration. *American Scientist, 86,* 236–244.

11 Gracco, V. L., & Abbs, J. H. (1986). Variant and invariant characteristics of speech movements. *Experimental Brain Research. Experimentelle Hirnforschung. Experimentation Cerebrale, 65,* 156–166.

12 Massaro, D. W. (1972). Preperceptual images, processing time, and perceptual units in auditory perception. *Psychological Review, 79,* 124–145.

13 Galantucci, B., Fowler, C. A., & Turvey, M. T. (2006). The motor theory of speech perception reviewed. *Psychonomic Bulletin & Review, 13,* 361–377.

14 Iacoboni, M. (2008). The role of premotor cortex in speech perception: Evidence from fMRI and rTMS. *Journal of Physiology, Paris, 102,* 31–34, doi:10.1016/j.jphysparis.2008.03.003; quotation from p. 32.

15 Fadiga, L., Craighero, L., Buccino, G., & Rizzolatti, G. (2002). Speech listening specifically modulates the excitability of tongue muscles: A TMS study. *The European Journal of Neuroscience, 15,* 399–402, doi:1874 [pii]; quotation from p. 399.

16 Fadiga, Craighero, Buccino, & Rizzolatti (2002), Speech listening specifically modulates the excitability of tongue muscles.

17 D'Ausilio, A., et al. (2009). The motor somatotopy of speech perception. *Current Biology, 19,* 381–385. doi:S0960-9822(09)00556-9 [pii] 10.1016/j.cub.2009.01.017.

18 Venezia, J. H., Saberi, K., Chubb, C., & Hickok, G. (2012). Response bias modulates the speech motor system during syllable discrimination. *Frontiers in Psychology, 3,* doi:10.3389/fpsyg.2012.00157.

19 Poeppel, D. (2001). Pure word deafness and the bilateral processing of the speech code. *Cognitive Science, 25,* 679–693.

20 Geschwind, N. (1965). Disconnexion syndromes in animals and man. *Brain, 88,* 237–294, 585–644.

21 Ross, E. D. (1981). The aprosodias: Functional-anatomic organization of the affective components of language in the right hemisphere. *Archives of Neurology, 38,* 561–569.

22   Damasio, H., & Damasio, A. R. (1980). The anatomical basis of conduction aphasia. *Brain, 103*, 337–350.

23   Baker, E., Blumstein, S. E., & Goodglass, H. (1981). Interaction between phonological and semantic factors in auditory comprehension. *Neuropsychologia, 19*, 1–15.

24   Basso, A., Casati, G., & Vignolo, L. A. (1977). Phonemic identification defects in aphasia. *Cortex, 13*, 84–95.

25   Miceli, G., Gainotti, G., Caltagirone, C., & Masullo, C. (1980). Some aspects of phonological impairment in aphasia. *Brain and Language, 11*, 159–169.

26   Hickok & Poeppel (2007), The cortical organization of speech processing; Hickok, G., & Poeppel, D. (2000). Towards a functional neuroanatomy of speech perception. *Trends in Cognitive Sciences, 4*, 131–138; and Hickok, G., & Poeppel, D. (2004). Dorsal and ventral streams: A framework for understanding aspects of the functional anatomy of language. *Cognition, 92*, 67–99.

27   Morais, J., Bertelson, P., Cary, L., & Alegria, J. (1986). Literacy training and speech segmentation. *Cognition, 24*, 45–64.

28   Mottonen, R., & Watkins, K. E. (2009). Motor representations of articulators contribute to categorical perception of speech sounds. *The Journal of Neuroscience: The Official Journal of the Society for Neuroscience, 29*, 9819–9825, doi:29/31/9819 [pii] 10.1523/JNEUROSCI.6018-08.2009; Meister, I. G., Wilson, S. M., Deblieck, C., Wu, A. D., & Iacoboni, M. (2007). The essential role of premotor cortex in speech perception. *Current Biology, 17*, 1692–1696, doi:S0960-9822(07)01969-0 [pii]10.1016/j.cub.2007.08.064.

29   Pulvermuller, F., et al. (2006). Motor cortex maps articulatory features of speech sounds. *Procedings of the National Academy of Sciences of the United States of America, 103*, 7865–7870, doi:10.1073/pnas.0509989103; and Wilson, S. M., Saygin, A. P., Sereno, M. I., & Iacoboni, M. (2004). Listening to speech activates motor areas involved in speech production. *Nature Neuroscience, 7*, 701–702.

30   See note 17 for full citation.

31   See note 28 for full citation.

32   See note 28 for full citation.

33   Lenneberg, E. H. (1962). Understanding language without ability to speak: A case report. *Journal of Abnormal and Social Psychology, 65*, 419–425.

34   Weller, M. (1993). Anterior opercular cortex lesions cause dissociated lower cranial nerve palsies and anarthria but no aphasia: Foix-Chavany-Marie syndrome and "automatic voluntary dissociation" revisited. *Journal of Neurology, 240*, 199–208.

35   Bishop, D. V., Brown, B. B., & Robson, J. (1990). The relationship between phoneme discrimination, speech production, and language comprehension in cerebral-palsied individuals. *Journal of Speech and Hearing Research, 33*, 210–219.

36   Rogalsky, C., Raphel, K., Poppa, T., Anderson, S., Damasio, H., Love, T., & Hickok, G. (2013). The neural basis of speech perception is task-dependent and does not rely on the motor system: A lesion study. Paper presented at the Cognitive Neuroscience Society.

37   Hickok, G., et al. (2008). Bilateral capacity for speech sound processing in

auditory comprehension: Evidence from Wada procedures. *Brain and Language, 107,* 179–184, doi:S0093-934X(08)00126-0 [pii] 10.1016/j.bandl.2008.09.006.

CHAPTER 6: *The Embodied Brain*

1   Skinner, B. F. (1974). *About Behaviorism.* New York: Alfred A. Knopf; quotation from pp. 17–18.

2   Chomsky, N., (1959). Verbal behavior by B.F. Skinner. *Language, 35,* 26–58; quotation from pp. 31–32.

3   Fodor, J. (1975). *The language of thought.* Cambridge, MA: MIT Press; Marr, D. (1981). *Vision: A computational investigation into the human representation and processing of visual information.* Cambridge, MA: MIT Press; and Pinker, S. (1997). *How the mind works.* New York: Norton.

4   Newell, A., Shaw, J. C., & Simon, H. A. (1958). Elements of a theory of human problem solving. *Psychological Review, 65,* 151–166; quotation from p. 153.

5   Recanzone, G. H., & Sutter, M. L. (2008). The biological basis of audition. *Annual Review of Psychology, 59,* 119–142, doi:10.1146/annurev.psych.59.103006.093544.

6   Neisser, U. (1967). *Cognitive psychology.* New York: Appleton-Century-Crofts; quotation from p. 40.

7   Gollisch, T., & Meister, M. (2010). Eye smarter than scientists believed: Neural computations in circuits of the retina. *Neuron, 65,* 150–164, doi:10.1016/j .neuron.2009.12.009.

8   Wolpert, D. M., Miall, R. C., & Kawato, M. (1998). Internal models in the cerebellum. *Trends in Cognitve Sciences, 2,* 338–347.

9   Neisser (1967), *Cognitive psychology.*

10   Engel, A. K., Maye, A., Kurthen, M., & Konig, P. (2013). Where's the action? The pragmatic turn in cognitive science. *Trends in Cognitive Sciences, 17,* 202–209. doi:10.1016/j.tics.2013.03.006.

11   Gomila, T., & Calvo, P. (2008). In P. Calvo & T. Gomila (Eds). *Handbook of cognitive science: An embodied approach* (pp.1–25). San Diego: Elsevier.

12   Hurley, S. (2008). The shared circuits model (SCM): How control, mirroring, and simulation can enable imitation, deliberation, and mind reading. *The Behavioral and Brain Sciences, 31,* 1–22, quotation from p. 2; discussion 22–58, doi:10.1017/S0140525X07003123.

13   Pezzulo, G., et al. (2011). The mechanics of embodiment: A dialog on embodiment and computational modeling. *Frontiers in Psychology, 2,* 5; quotation from p. 3. doi:10.3389/fpsyg.2011.00005.

14   Gomila & Calvo (2008), *Handbook of cognitive science, 7.*

15   Gomila & Calvo (2008), *Handbook of cognitive science, 8.*

16   Pylyshyn, Z. W. (1981). The imagery debate: Analogue media versus tacit knowledge. *Psychological Review, 88,* 16–45.

17   Smith, E. E. & Medin, D. L. (1981). *Categories and concepts.* Cambridge, MA: Harvard University Press.

18   Warrington, E. K., & Shallice, T. (1984). Category specific semantic impairments. *Brain, 107 (Pt 3),* 829–854.

19   Gainotti, G., Silveri, M. C., Daniele, A., & Giustolisi, L. (1995). Neuroanatomical correlates of category-specific semantic disorders: a critical survey. *Memory, 3,* 247–264, doi:10.1080/09658219508253153; and Tranel, D., Kemmerer, D., Adolphs, R., Damasio, H., & Damasio, A. R. (2003). Neural correlates of conceptual knowledge for actions. *Cognitive Neuropsychology, 20,* 409–432.

20   Shepard, R. N., & Metzler, J. (1971). Mental rotation of three-dimensional objects. *Science, 171,* 701–703; and Kosslyn, S. M. (1996). *Image and brain: The resolution of the imagery debate.* Cambridge, MA: MIT Press.

21   Arora, S., et al. (2011). Mental practice enhances surgical technical skills: A randomized controlled study. *Annals of Surgery, 253,* 265–270, doi:10.1097/SLA.0b013e318207a789.

22   Barsalou, L. W. (2012). The human conceptual system. In M. J. Spivey, K. McRae, & M. F. Joanisse (Eds.), *The Cambridge handbook of psycholinguistics* (pp. 239–258). New York: Cambridge University Press; quotation from p. 251.

23   Oztop, E., Kawato, M., & Arbib, M. (2006). Mirror neurons and imitation: A computationally guided review. *Neural Networks, 19,* 254–271, doi:10.1016/j.neunet.2006.02.002.

24   Rizzolatti, G., Fogassi, L., & Gallese, V. (2001). Neurophysiological mechanisms underlying the understanding and imitation of action. *Nature Reviews Neuroscience, 2,* 661–670.

25   Gallese (2001), The "shared manifold" hypothesis; and Gallese, V., & Sinigaglia, C. (2011). What is so special about embodied simulation? *Trends in Cognitive Sciences, 15,* 512–519, doi:10.1016/j.tics.2011.09.003.

26   Keysers, C., et al. (2004). A touching sight: SII/PV activation during the observation and experience of touch. *Neuron, 42,* 335–346; Saarela, M. V., et al. (2007). The compassionate brain: Humans detect intensity of pain from another's face. *Cerebral Cortex, 17,* 230–237, doi:10.1093/cercor/bhj141; and Wicker, B., et al. (2003). Both of us disgusted in My insula: The common neural basis of seeing and feeling disgust. *Neuron, 40,* 655–664.

27   Hauk, Johnsrude, & Pulvermuller (2004), Somatotopic representation of action words.

28   Glenberg, A. M., & Kaschak, M. P. (2002). Grounding language in action. *Psychonomic Bulletin & Review, 9,* 558–565.

29   Gallese, Fadiga, Fogassi, & Rizzolatti (1996), Action recognition in the premotor cortex.

30   Geyer, S., Matelli, M., Luppino, G., & Zilles, K. (2000). Functional neuroanatomy of the primate isocortical motor system. *Anatomy and Embryology, 202,* 443–474; and Georgopoulos, A. P., Kalaska, J. F., Caminiti, R., & Massey, J. T. (1982). On the relations between the direction of two-dimensional arm movements and cell discharge in primate motor cortex. *Journal of Neuroscience, 2,* 1527–1537.

31   Mahon, B. Z., & Caramazza, A. (2008). A critical look at the embodied cognition hypothesis and a new proposal for grounding conceptual content. *Journal of Physiology, Paris, 102,* 59–70; see also Hickok, G. (2010). The role of mirror neurons

in speech perception and action word semantics. *Language and Cognitive Processes, 25,* 749–776.

32 de Zubicaray, G., Arciuli, J., & McMahon, K. (2013). Putting an "end" to the motor cortex representations of action words. *Journal of Cognitive Neuroscience, 25,* 1957–1974. doi:10.1162/jocn_a_00437.

33 Watson, C. E., Cardillo, E. R., Ianni, G. R., & Chatterjee, A. (2013). Action concepts in the brain: An activation likelihood estimation meta-analysis. *Journal of Cognitive Neuroscience, 25,* 1191–1205. doi:10.1162/jocn_a_00401.

34 Pinker, S. (2007). *The stuff of thought: Language as a window into human nature.* New York: Viking.

35 Bak, T. H., O'Donovan, D. G., Xuereb, J. H., Boniface, S., & Hodges, J. R. (2001). Selective impairment of verb processing associated with pathological changes in Brodmann areas 44 and 45 in the motor neurone disease–dementia–aphasia syndrome. *Brain, 124,* 103–120.

36 Glenberg, A. M., & Gallese, V. (2012). Action-based language: a theory of language acquisition, comprehension, and production. *Cortex, 48,* 905–922. doi:10.1016/j.cortex.2011.04.010.

37 Bak, O'Donovan, Xuereb, Boniface, & Hodges (2001), Selective impairment of verb processing.

38 Bak, O'Donovan, Xuereb, Boniface, & Hodges (2001), Selective impairment of verb processing.

39 Hickok (2010), The role of mirror neurons in speech perception; and Novick, J. M., Trueswell, J. C., & Thompson-Schill, S. L. (2005). Cognitive control and parsing: Reexamining the role of Broca's area in sentence comprehension. *Cognitive, Affective & Behavioral Neuroscience, 5,* 263–281.

40 Grossman, M., et al. (2008). Impaired action knowledge in amyotrophic lateral sclerosis. *Neurology, 71,* 1396–1401. doi:10.1212/01.wnl.0000319701.50168.8c.

41 Bono. (n.d.). Retrieved from http://www.u2star.com/blog/discography/how -to-dismantle-an-atomic-bomb/miracle-drug

42 Grimes, W. (2009, February 24). Christopher Nolan, Irish author, dies at 43. *The New York Times,* p. A18.

43 n.a. (2000). Retrieved from http://www.publishersweekly.com/978-1-55970 -511-0

44 Geracimos, A. (2000). Retrieved from http://www.publishersweekly.com/ pw/by-topic/authors/interviews/article/19743-pw-christopher-nolan-against-all- odds.html

45 Tranel, D., Adolphs, R., Damasio, H., & Damasio, A. R. (2001). A neural basis for the retrieval of words for actions. *Cognitive Neuropsychology, 18,* 655–670.

46 Hillis, A. E., Oh, S., & Ken, L. (2004). Deterioration of naming nouns versus verbs in primary progressive aphasia. *Annals of Neurology, 55,* 268–275. doi:10.1002/ ana.10812.

47 Hillis, A. E., Tuffiash, E., Wityk, R. J., & Barker, P. B. (2002). Regions of neural dysfunction associated with impaired naming of actions and objects in acute stroke. *Cognitive Neuropsychology, 19,* 523–534.

48 Willems, R. M., Labruna, L., D'Esposito, M., Ivry, R., & Casasanto, D. (2011).

A functional role for the motor system in language understanding: Evidence from theta-burst transcranial magnetic stimulation. *Psychological Science, 22,* 849–854, doi:10.1177/0956797611412387.

49 Hecht, H., Vogt, S., & Prinz, W. (2001). Motor learning enhances perceptual judgment: A case for action-perception transfer. *Psychological Research, 65,* 3–14; and James, K. H., et al. (2002). "Active" and "passive" learning of three-dimensional object structure within an immersive virtual reality environment. *Behavior Research Methods, Instruments, & Computers: A Journal of the Psychonomic Society, Inc, 34,* 383–390.

50 Rizzolatti & Sinigaglia (2010), The functional role of the parieto-frontal mirror circuit.

51 Oztop, Kawato, & Arbib (2006), Mirror neurons and imitation.

CHAPTER 7: *Feeling, Doing, Knowing*

1 Cole, J. D., & Sedgwick, E. M. (1992). The perceptions of force and of movement in a man without large myelinated sensory afferents below the neck. *The Journal of Physiology, 449,* 503–515.

2 Bastian, H. C. (1887). The "muscular sense": Its nature and cortical localisation. *Brain, 10,* 1–89.

3 Sanes, J. N., Mauritz, K. H., Evarts, E. V., Dalakas, M. C., & Chu, A. (1984). Motor deficits in patients with large-fiber sensory neuropathy. *Proceedings of the National Academy of Sciences of the United States of America, 81,* 979–982.

4 Smith, C. R. (1975). Residual hearing and speech production in deaf children. *Journal of Speech and Hearing Research, 18,* 795–811.

5 Sancier, M. L., & Fowler, C. A. (1997). Gestural drift in a bilingual speaker of Brazilian Portuguese and English. *Journal of Phonetics, 25,* 421–436.

6 Davis, H. (2010, October 6). Can you say something without turning it into a question? Retrieved from http://www.psychologytoday.com/blog/caveman-logic/201010/the-uptalk-epidemic.

7 Gorman, J. (1993, August 15). On language; Like, Uptalk? *The New York Times.* Retrieved from http://www.nytimes.com/1993/08/15/magazine/on-language-like-uptalk.html.

8 Liberman, M. (2006, March 28). Uptalk is not HRT. Retrieved from http://itre.cis.upenn.edu/~myl/languagelog/archives/002967.html.

9 Gibson, J. J. (1979). *The ecological approach to visual perception.* Boston: Houghton Mifflin.

10 Gallese, V., Gernsbacher, M. A., Heyes, C., Hickok, G., & Iacoboni, M. (2011). Mirror neuron forum. *Perspectives on Psychological Science, 6,* 369–407.

11 Liberman, A. M., & Mattingly, I. G. (1985). The motor theory of speech perception revised. *Cognition, 21,* 1–36.

12 Guenther, F. H., Hampson, M., & Johnson, D. (1998). A theoretical investigation of reference frames for the planning of speech movements. *Psychological Review, 105,* 611–633.

13   Gallese, Gernsbacher, Heyes, Hickok, & Iacoboni (2011), Mirror neuron forum.

14   Glenberg & Gallese (2012), Action-based language.

15   Chouinard, P. A., & Paus, T. (2006). The primary motor and premotor areas of the human cerebral cortex. *The Neuroscientist, 12,* 143–152, doi:10.1177 /1073858405284255; and Felleman, D. J., & Van Essen, D. C. (1991). Distributed hierarchical processing in primate visual cortex. *Cerebral Cortex, 1,* 1–47.

16   Felleman & Van Essen (1991), Distributed hierarchical processing in primate visual cortex.

17   Shadmehr, R., Smith, M. A., & Krakauer, J. W. (2010). Error correction, sensory prediction, and adaptation in motor control. *Annual Review of Neuroscience, 33,* 89–108, doi:10.1146/annurev-neuro-060909-153135.

18   Milner & Goodale (1995), *The visual brain in action.*

19   Binder, J. R., Desai, R. H., Graves, W. W., & Conant, L. L. (2009). Where is the semantic system? A critical review and meta-analysis of 120 functional neuroimaging studies. *Cerebral Cortex, 19,* 2767–2796. doi:10.1093/cercor/bhp055.

20   Raichle, M. E. (2006). The brain's dark energy. *Science, 314,* 1249–1250. doi:10.1126/science. 1134405.

21   Cahn, B. R., & Polich, J. (2006). Meditation states and traits: EEG, ERP, and neuroimaging studies. *Psychological Bulletin, 132,* 180–211. doi:10.1037/0033 -2909.132.2.180.

22   Baron Short, E., et al. (2010). Regional brain activation during meditation shows time and practice effects: An exploratory fMRI study. *Evidence-Based Complementary and Alternative Medicine: eCAM, 7,* 121–127, doi:10.1093/ecam/nem163.

23   Binder, J. R., et al. (1999). Conceptual processing during the conscious resting state. A functional MRI study. *Journal of Cognitive Neuroscience, 11,* 80–95.

24   Raichle, M. E., et al. (2001). A default mode of brain function. *Proceedings of the National Academy of Sciences of the United States of America, 98,* 676–682. doi:10.1073/ pnas.98.2.676.

25   Buckner, R. L., Andrews-Hanna, J. R., & Schacter, D. L. (2008). The brain's default network: Anatomy, function, and relevance to disease. *Annals of the New York Academy of Sciences, 1124,* 1–38, doi:10.1196/annals.1440.011.

26   Buckner, Andrews-Hanna, & Schacter (2008), The brain's default network.

27   Buckner, Andrews-Hanna, & Schacter (2008), The brain's default network.

28   Martin, A. (2007). The representation of object concepts in the brain. *Annual Review of Psychology, 58,* 25–45. doi:10.1146/annurev.psych.57.102904.190143.

29   Martin (2007), The representation of object concepts in the brain; and Caramazza, A., & Shelton, J. R. (1998). Domain-specific knowledge systems in the brain the animate-inanimate distinction. *Journal of Cognitve Neuroscience, 10,* 1–34.

30   Hoffman, P., Pobric, G., Drakesmith, M., & Lambon Ralph, M. A. (2012). Posterior middle temporal gyrus is involved in verbal and non-verbal semantic cognition: Evidence from rTMS. *Aphasiology, 26,* 1119–1130.

31   Martin (2007), The representation of object concepts in the brain; Hodges, J. R., & Patterson, K. (2007). Semantic dementia: a unique clinicopathological syndrome. *Lancet Neurology, 6,* 1004–1014; and Patterson, K., Nestor, P. J., & Rogers,

T. T. (2007). Where do you know what you know? The representation of semantic knowledge in the human brain. *Nature Reviews Neuroscience, 8*, 976–987.

32 Gorno-Tempini, M. L., et al. (2011). Classification of primary progressive aphasia and its variants. *Neurology, 76*, 1006–1014. doi:WNL.0b013e31821103e6 [pii] 10.1212/WNL.0b013e31821103e6.

33 Hodges & Patterson (2007), Semantic dementia.

34 Acosta-Cabronero, J., et al. (2011). Atrophy, hypometabolism and white matter abnormalities in semantic dementia tell a coherent story. *Brain, 134*, 2025–2035. doi:10.1093/brain/awr119.

35 Pazzaglia, Smania, Corato, & Aglioti (2008), Neural underpinnings of gesture discrimination.

36 Nelissen, N., et al. (2010). Gesture discrimination in primary progressive aphasia: The intersection between gesture and language processing pathways. *The Journal of Neuroscience: The Official Journal of the Society for Neuroscience, 30*, 6334–6341. doi:30/18/6334 [pii] 10.1523/JNEUROSCI.0321-10.2010.

37 Kalenine, S., Buxbaum, L. J., & Coslett, H. B. (2010). Critical brain regions for action recognition: Lesion symptom mapping in left hemisphere stroke. *Brain, 133*, 3269–3280. doi:10.1093/brain/awq210.

38 Meligne, D., et al. (2011). Verb production during action naming in semantic dementia. *Journal of Communication Disorders, 44*, 379–391, doi:10.1016/j.jcomdis .2010.12.001; and Hoffman, P., Jones, R. W., & Lambon Ralph, M. A. (2012). Be concrete to be comprehended: Consistent imageability effects in semantic dementia for nouns, verbs, synonyms and associates. *Cortex, 49*, 1206–1218. doi:10.1016/j .cortex.2012.05.007.

39 Tranel, Kemmerer, Adolfs, Damasio, & Damasio (2003), Neural correlates of conceptual knowledge for actions.

40 Baddeley, A. D. (1992). Working memory. *Science, 255*, 556–559.

41 Fuster, J. M. (1995). *Memory in the cerebral cortex.* Cambridge, MA: MIT Press.

42 Damasio, A. R., & Damasio, H. (1994). Cortical systems for retrieval of concrete knowledge. In C. Koch & J. L. Davis (Eds.), *Large-scale neuronal theories of the brain* (pp. 61–74). Cambridge, MA: MIT Press.

43 Grossman, E., et al. (2000). Brain areas involved in perception of biological motion. *Journal of Cognitive Neuroscience, 12*, 711–720; and Grossman, E. D., Battelli, L., & Pascual-Leone, A. (2005). Repetitive TMS over posterior STS disrupts perception of biological motion. *Vision Research, 45*, 2847–2853, doi:10.1016/j .visres.2005.05.027.

44 Pelphrey, K. A., & Morris, J. P. (2006). Brain mechanisms for interpreting the actions of others from biological-motion cues. *Current Directions in Psychological Science, 15*, 136–140, doi:10.1111/j.0963-7214.2006.00423.x.

45 Pelphrey, K. A., Singerman, J. D., Allison, T., & McCarthy, G. (2003). Brain activation evoked by perception of gaze shifts: The influence of context. *Neuropsychologia, 41*, 156–170; Pelphrey, K. A., Viola, R. J., & McCarthy, G. (2004). When strangers pass: Processing of mutual and averted social gaze in the superior temporal sulcus. *Psychological Science, 15*, 598–603, doi:10.1111/j.0956-7976.2004.00726.x; and Shultz, S., Lee, S. M., Pelphrey, K., & McCarthy, G. (2011). The posterior supe-

rior temporal sulcus is sensitive to the outcome of human and non-human goal-directed actions. *Social Cognitive and Affective Neuroscience, 6,* 602–611, doi:10.1093/scan/nsq087.

46 Shultz, Lee, Pelphrey, & McCarthy (2011), The posterior superior temporal sulcus is sensitive.

47 Perrett et al. (1985), Visual analysis of body movements; and Perrett, Mistlin, Harries, and Chitty (1990), *Vision and action.*

48 Perrett, D. I., et al. (1985). Visual cells in the temporal cortex sensitive to face view and gaze direction. *Proceedings of the Royal Society of London B Biological Sciences, 223,* 293–317.

49 Jastorff, J., Popivanov, I. D., Vogels, R., Vanduffel, W., & Orban, G. A. (2012). Integration of shape and motion cues in biological motion processing in the monkey STS. *Neuroimage, 60,* 911–921. doi:10.1016/j.neuroimage.2011.12.087.

50 Oztop, E., & Arbib, M. A. (2002). Schema design and implementation of the grasp-related mirror neuron system. *Biological Cybernetics, 87,* 116–140. doi:10.1007/s00422-002-0318-1.

51 Oztop, Kawato, & Arbib (2006), Mirror neurons and imitation.

CHAPTER 8: *Human Imitans and the Function of Mirror Neurons*

1 Meltzoff, A. N., & Moore, M. K. (1983). Newborn infants imitate adult facial gestures. *Child development, 54,* 702–709.

2 Meltzoff, A. N. (1985). Immediate and deferred imitation in fourteen- and twenty-four-month-old infants. *Child Development, 56,* 62–72.

3 Kuhl, P. K., & Meltzoff, A. N. (1982). The bimodal perception of speech in infancy. *Science, 218,* 1138–1141.

4 Meltzoff, A. N., & Moore, M. K. (1994). Imitation, memory, and the representation of persons. *Infant Behavior and Development, 17,* 83–99.

5 Carpenter, M., Akhtar, N., & Tomasello, M. (1998). Fourteen- through 18-month-old infants differentially imitate intentional and accidental actions. *Infant Behavior and Development, 21,* 315–330.

6 Meltzoff, A. N. (1995). Understanding the intentions of others: Re-enactment of intended acts by 18-month-old children. *Developmental Psychology, 31,* 838–850.

7 Meltzoff, A. N., Kuhl, P. K., Movellan, J., & Sejnowski, T. J. (2009). Foundations for a new science of learning. *Science, 325,* 284–288. doi:10.1126/science.1175626.

8 Hanna, E., & Meltzoff, A. (1993). Peer imitation by toddlers in laboratory, home, and day-care contexts: Implications for social learning and memory. *Developmental Psychology, 29,* 701–710.

9 Hogberg, A. (2008). Playing with flint: Tracing a child's imitation of adult work in a lithic assemblage. *Journal of Archaeological Method and Theory, 15,* 112–131.

10 Meltzoff, A. N., & Decety, J. (2003). What imitation tells us about social cognition: A rapprochement between developmental psychology and cognitive neuroscience. *Philosophical Transactions of the Royal Society of London. Series B, Biological Sciences, 358,* 491–500. doi:10.1098/rstb.2002.1261.

11 Heyes, C. (2013). What can imitation do for cooperation? In K. Sterelny, R. Joyce, B. Calcott, & B. Fraser (Eds.), *Cooperation and its evolution* (pp. 313–332). Cambridge, MA: MIT Press.

12 Jeannerod, M. (1994). The representing brain: Neural correlates of motor intention and imagery. *Behavioral and Brain Sciences, 17*, 187–202.

13 Rizzolatti & Craighero (2004), The mirror-neuron system, 172.

14 Meltzoff & Decety (2003), What imitation tells us about social cognition, 494.

15 Oztop, Kawato, & Arbib (2006), Mirror neurons and imitation.

16 Rizzolatti & Craighero (2004), The mirror-neuron system.

17 Rizzolatti & Craighero (2004), The mirror-neuron system, 183.

18 Rizzolatti & Craighero (2004), The mirror-neuron system; and Iacoboni et al. (1999), Cortical mechanisms of human communication.

19 di Pellegrino, Fadiga, Fogassi, Gallese, & Rizzolatti (1992), Understanding motor events, 179.

20 Heyes (2013), in *Cooperation and its evolution*.

21 Subiaul, F., Cantlon, J. F., Holloway, R. L., & Terrace, H. S. (2004). Cognitive imitation in rhesus macaques. *Science, 305*, 407–410. doi:10.1126/science.1099136.

22 Subiaul, F., Romansky, K., Cantlon, J. F., Klein, T., & Terrace, H. (2007). Cognitive imitation in 2-year-old children (*Homo sapiens*): A comparison with rhesus monkeys (*Macaca mulatta*). *Animal Cognition, 10*, 369–375. doi:10.1007/s10071-006-0070-3.

23 Huffman, M. A., Nahallage, C. A. D., & Leca, J.-B. (2008). Cultured monkeys: Social learning cast in stones. *Current Directions in Psychological Science, 17*, 410–414.

24 Voelkl, B., & Huber, L. (2007). Imitation as faithful copying of a novel technique in marmoset monkeys. *PLoS One, 2*, e611. doi:10.1371/journal.pone.0000611.

25 Range, F., Viranyi, Z., & Huber, L. (2007). Selective imitation in domestic dogs. *Current Biology, 17*, 868–872. doi:S0960-9822(07)01267-5 [pii] 10.1016/j.cub.2007.04.026.

26 Muller, C. A., & Cant, M. A. (2010). Imitation and traditions in wild banded mongooses. *Current Biology, 20*, 1171–1175. doi:S0960-9822(10)00514-2 [pii] 10.1016/j.cub.2010.04.037.

27 Krutzen, M., et al. (2005). Cultural transmission of tool use in bottlenose dolphins. *Proceedings of the National Academy of Sciences of the United States of America, 102*, 8939–8943. doi:0500232102 [pii] 10.1073/pnas.0500232102.

28 Page, R. A., & Ryan, M. J. (2006). Social transmission of novel foraging behavior in bats: Frog calls and their referents. *Current Biology, 16*, 1201–1205, doi:S0960-9822(06)01499-0 [pii] 10.1016/j.cub.2006.04.038.

29 Schuster, S., Wohl, S., Griebsch, M., & Klostermeier, I. (2006). Animal cognition: How archer fish learn to down rapidly moving targets. *Current Biology, 16*, 378–383. doi:S0960-9822(06)01013-X [pii] 10.1016/j.cub.2005.12.037.

30 Fiorito, G., & Scotto, P. (1992). Observational learning in *Octopus vulgaris*. *Science, 256*, 545–547. doi:256/5056/545 [pii] 10.1126/science.256.5056.545.

31 Oztop & Arbib (2002), Schema design and implementation of the grasp-related mirror neuron system.

32 Heyes, C. (2010). Where do mirror neurons come from? *Neuroscience and Biobe-*

havioral Reviews, 34, 575–583. doi:S0149-7634(09)00173-0 [pii] 10.1016/j.neubiorev .2009.11.007.

33   Oztop, Kawato, & Arbib (2006), Mirror neurons and imitation.

34   Gallese, Fadiga, Fogassi, & Rizzolatti (1996), Action recognition in the premotor cortex.

35   Umilta et al. (2001), I know what you are doing.

36   Ferrari, Rozzi, & Fogassi (2005), Mirror neurons responding to observation of actions.

37   di Pellegrino, Fadiga, Fogassi, Gallese, & Rizzolatti (1992), Understanding motor events, 179.

38   Meltzoff (1995), Understanding the intentions of others.

39   Berko, J. (1958). The child's learning of English morphology. *Word, 14,* 150–177.

40   Zajonc, R. B., Adelmann, P. K., Murphy, S. T., & Niedenthal, P. M. (1987). Convergence in the physical appearance of spouses. *Motivation and Emotion, 11,* 335–346.

41   Chartrand, T. L., & Bargh, J. A. (1999). The chameleon effect: The perception-behavior link and social interaction. *Journal of Personality and Social Psychology, 76,* 893–910.

42   van Baaren, R., Janssen, L., Chartrand, T. L., & Dijksterhuis, A. (2009). Where is the love? The social aspects of mimicry. *Philosophical Transactions of the Royal Society of London. Series B, Biological Sciences, 364,* 2381–2389. doi:10.1098/rstb.2009.0057.

43   Ganos, C., Ogrzal, T., Schnitzler, A., & Munchau, A. (2012). The pathophysiology of echopraxia/echolalia: Relevance to Gilles de la Tourette syndrome. *Movement Disorders: Official Journal of the Movement Disorder Society, 27,* 1222–1229. doi:10.1002/mds.25103.

44   van Baaren, Janssen, Chartrand, & Dijksterhuis (2009), Where is the love?

45   Chartrand & Bargh (1999), The chameleon effect, 238.

46   Jones, S. S. (2009). The development of imitation in infancy. *Philosophical Transactions of the Royal Society of London. Series B, Biological Sciences, 364,* 2325–2335; quotation from p. 2325. doi:10.1098/rstb.2009.0045.

CHAPTER 9: *Broken Mirrors*

1   Williams, Whiten, Suddendorf, & Perrett (2001), Imitation, mirror neurons, and autism; Gallese, V. (2006). Intentional attunement: A neurophysiological perspective on social cognition and its disruption in autism. *Brain Research, 1079,* 15–24; Baron-Cohen, S. (1995). *Mindblindness: An essay on autism and theory of mind.* Cambridge, MA: MIT Press; Baron-Cohen, S., Leslie, A. M., & Frith, U. (1985). Does the autistic child have a "theory of mind"? *Cognition, 21,* 37–46; and Gernsbacher, M. A., & Frymiare, J. L. (2005). Does the autistic brain lack core modules? *The Journal of Developmental and Learning Disorders, 9,* 3–16.

2   American Psychiatric Association. (2013). Diagnostic and statistical manual of mental disorders (5th ed.). Arlington, VA: American Psychiatric Publishing.

3   Tordjman, S., et al. (2009). Pain reactivity and plasma beta-endorphin in chil-

dren and adolescents with autistic disorder. *PLoS One, 4,* e5289, doi:10.1371/journal.pone.0005289.

4   Baron-Cohen, Leslie, & Firth (1985), Does the autistic child have a "theory of mind"?

5   Baron-Cohen (1995), *Mindblindness.*

6   Gallese (2006), Intentional attunement, 21–22.

7   Gallese (2006), Intentional attunement, 21.

8   Williams, Whiten, Suddendorf, & Perrett (2001), Imitation, mirror neurons, and autism; and Oberman, L. M., et al. (2005). EEG evidence for mirror neuron dysfunction in autism spectrum disorders. *Cognitive Brain Research, 24,* 190–198.

9   Hamilton, A. F. (2009). Goals, intentions and mental states: Challenges for theories of autism. *Journal of Child Psychology and Psychiatry, and Allied Disciplines, 50,* 881–892. doi:10.1111/j.1469-7610.2009.02098.x.

10   Hamilton, A. F., Brindley, R. M., & Frith, U. (2007). Imitation and action understanding in autistic spectrum disorders: How valid is the hypothesis of a deficit in the mirror neuron system? *Neuropsychologia, 45,* 1859–1868, doi:10.1016/j.neuropsychologia.2006.11.022.

11   Gallese, Gernsbacher, Heyes, Hickok, & Iacoboni (2011), Mirror neuron forum, 390.

12   Oberman, et al. (2005), EEG evidence for mirror neuron dysfunction in autism spectrum disorders; and Dapretto, M., et al. (2006). Understanding emotions in others: Mirror neuron dysfunction in children with autism spectrum disorders. *Nature Neuroscience, 9,* 28–30, doi:10.1038/nn1611.

13   Gallese, Gernsbacher, Heyes, Hickok, & Iacoboni (2011), Mirror neuron forum

14   Gallese, Gernsbacher, Heyes, Hickok, & Iacoboni (2011), Mirror neuron forum; and Bird, G., Leighton, J., Press, C., & Heyes, C. (2007). Intact automatic imitation of human and robot actions in autism spectrum disorders. *Proceedings of the Royal Society of London B Biological Sciences, 274,* 3027–3031. doi:10.1098/rspb.2007.1019.

15   Bloom, P., & German, T. P. (2000). Two reasons to abandon the false-belief task as a test of theory of mind. *Cognition, 77,* B25–31.

16   Gernsbacher & Frymiare (2005). Does the autistic brain lack core modules? *Journal of Developmental and Learning Disorders 9,* 3–16; and Moran, J. M., et al. (2011). Impaired theory of mind for moral judgment in high-functioning autism. *Proceedings of the National Academy of Sciences of the United States of America, 108,* 2688-2692. doi:10.1073/pnas.1011734108.

17   Gernsbacher & Frymiare (2005), Does the autistic brain lack core modules?

18   Bloom & German (2000), Two reasons to abandon the false belief task.

19   Peterson, C. C., Wellman, H. M., & Liu, D. (2005). Steps in theory-of-mind development for children with deafness or autism. *Child Development, 76,* 502–517, doi:10.1111/j.1467-8624.2005.00859.x.

20   Hamilton (2009), Goals, intentions and mental states, 888.

21   Markram, H., Rinaldi, T., & Markram, K. (2007). The intense world syndrome—an alternative hypothesis for autism. *Frontiers in Neuroscience, 1,* 77–96. doi:10.3389/neuro.01.1.1.006.2007.

22   Markram, K., & Markram, H. (2010) The intense world theory—a unifying theory of the neurobiology of autism. *Frontiers in Human Neuroscience, 4,* 224. doi:10.3389/fnhum.2010.00224.

23   Markram & Markram (2010), The intense world theory, 3.

24   Marco, E. J., Hinkley, L. B., Hill, S. S., & Nagarajan, S. S. (2011). Sensory processing in autism: A review of neurophysiologic findings. *Pediatric Research, 69,* 48R–54R. doi:10.1203/PDR.0b013e3182130c54.

25   Grandin, T. (2006). Retrieved from http://www.npr.org/templates/story/story .php?storyId=5488844.

26   Dalton, K. M., et al. (2005). Gaze fixation and the neural circuitry of face processing in autism. *Nature Neuroscience, 8,* 519–526, doi:10.1038/nn1421.

27   Gernsbacher & Frymiare (2005). Does the autistic brain lack core modules?

CHAPTER 10: *Predicting the Future of Mirror Neurons*

1   Rizzolatti, Fogassi, & Gallese (2001), Neurophysiological mechanisms.

2   Rizzolatti, Fogassi, & Gallese (2001), Neurophysiological mechanisms.

3   Csibra (2007), Sensorimotor foundations of higher cognition.

4   Arbib, M. A. (2004). *Beyond the mirror.* Oxford University Press.

5   Csibra (2007), Sensorimotor foundations of higher cognition.

6   Heyes (2010), Where do mirror neurons come from?

7   Hurford, J. (2004). Language beyond our grasp: What mirror neurons can, and cannot, do for language evolution. In D. Kimbrough Oller & U. Griebel (Eds.), *Evolution of communication systems: A comparative approach* (pp. 297–313). Cambridge, MA: MIT Press.

8   Jacob, P., & Jeannerod, M. (2005). The motor theory of social cognition: A critique. *Trends in Cognitive Sciences, 9,* 21–25. doi:10.1016/j.tics.2004.11.003.

9   Kilner, J. M. (2011). More than one pathway to action understanding. *Trends in Cognitive Sciences, 15,* 352–357, doi:10.1016/j.tics.2011.06.005; and Kilner, J. M., Friston, K. J., & Frith, C. D. (2007). Predictive coding: An account of the mirror neuron system. *Cognitive processing, 8,* 159–166, doi:10.1007/s10339-007-0170-2.

10   Oztop & Arbib (2002), Schema design and implementation of the grasp-related mirror neuron system.

11   Kilner (2011), More than one pathway to action understanding; and Kilner, Friston, and Frith (2007), Predictive coding.

12   Kawato, M. (1999). Internal models for motor control and trajectory planning. *Current Opinion in Neurobiology, 9,* 718–727; Shadmehr, R., & Krakauer, J. W. (2008). A computational neuroanatomy for motor control. *Experimental Brain Research. Experimentelle Hirnforschung. Experimentation Cerebrale, 185,* 359–381, doi:10.1007/ s00221-008-1280-5; and Wolpert, D. M., Ghahramani, Z., & Jordan, M. I. (1995). An internal model for sensorimotor integration. *Science, 269,* 1880–1882.

13   Blakemore, S. J., Wolpert, D., & Frith, C. (2000). Why can't you tickle yourself? *Neuroreport, 11,* R11–16.

14   Bremmer, F., Kubischik, M., Hoffmann, K. P., & Krekelberg, B. (2009). Neural

dynamics of saccadic suppression. *The Journal of Neuroscience: The Official Journal of the Society for Neuroscience, 29,* 12374–12383. doi:10.1523/JNEUROSCI.2908-09.2009.

15   Houde, J. F., Nagarajan, S. S., Sekihara, K., & Merzenich, M. M. (2002). Modulation of the auditory cortex during speech: An MEG study. *Journal of Cognitive Neuroscience, 14,* 1125–1138. doi:10.1162/089892902760807140.

16   Eliades, S. J., & Wang, X. (2008). Neural substrates of vocalization feedback monitoring in primate auditory cortex. *Nature, 453,* 1102–1106. doi:10.1038/nature06910; and Eliades, S. J., & Wang, X. (2013). Comparison of auditory-vocal interactions across multiple types of vocalizations in marmoset auditory cortex. *Journal of Neurophysiology, 109,* 1638–1657, doi:10.1152/jn.00698.2012.

17   Stevens, K. N., & Halle, M. (1967). Remarks on the analysis by synthesis and distinctive features. In W. Walthen-Dunn (Ed.), *Models for the perception of speech and visual form* (pp. 88–102). Cambridge, MA: MIT Press; Wilson, S. M., & Iacoboni, M.(2006). Neural responses to non-native phonemes varying in producibility: Evidence for the sensorimotor nature of speech perception. *Neuroimage, 33,* 316–325, doi:10.1016/j.neuroimage.2006.05.032; and Bever, T. G., & Poeppel, D. (2010). Analysis by synthesis: A (re-)emerging program of research for language and vision. *Biolinguistics, 4.2-3,* 174–200.

18   Hickok, G., Houde, J., & Rong, F. (2011). Sensorimotor integration in speech processing: Computational basis and neural organization. *Neuron, 69,* 407–422, doi:S0896-6273(11)00067-5 [pii]10.1016/j.neuron.2011.01.019.

19   James, W. (1890). *The principles of psychology* (p. 300) Vol. 2. New York: Henry Holt and Company.

20   Salin, P. A., & Bullier, J. (1995). Corticocortical connections in the visual system: Structure and function. *Physiological Reviews, 75,* 107–154; and Muckli, L., & Petro, L. S. (2013). Network interactions: Non-geniculate input to V1. *Current Opinion in Neurobiology, 23,* 195–201, doi:10.1016/j.conb.2013.01.020.

21   Friston, K. (2010). The free-energy principle: A unified brain theory? *Nature Reviews Neuroscience, 11,* 127-138, doi:10.1038/nrn2787; and Friston, K. J., Daunizeau, J., Kilner, J., & Kiebel, S. J. (2010). Action and behavior: A free-energy formulation. *Biological Cybernetics, 102,* 227–260, doi:10.1007/s00422-010-0364-z.

22   Fodor, J. A. (1983). *The modularity of mind.* Cambridge, MA: MIT Press.

Index

Page numbers in *italics* refer to illustrations.